FLORA OF TROPICAL EAST AFRICA

MELIACEAE

B.T. Styles & F. White

Department of Plant Sciences, University of Oxford

Trees, shrubs or shrublets. Wood often scented. Indumentum of simple, glandular or stellate hairs. Leaves usually alternate, 2- or 3-pinnate, simply impari- or paripinnate, 3-foliolate, 1-foliolate or simple; leaflets entire, crenate or serrate. Stipules absent. Inflorescence usually axillary or in axils of fallen leaves, of cymose panicles or of compound or simple cymes or flowers fasciculate, rarely solitary. Flowers bisexual or unisexual, monoecious or dioecious, occasionally polygamous, actinomorphic, mainly 4–5-merous; sepals and petals dissimilar. Sepals small, 4–6, variously connate or almost free, the lobes imbricate or with open aestivation, never completely covering corolla in bud. Petals usually 4–5, free (adnate to the staminal tube in *Turraeanthus*), valvate, imbricate or contorted. Disk intrastaminal, very variable, rarely absent, often developed from the gynophore, completely fused to base of staminal tube, or annular, cup-shaped or cushion-shaped and free from staminal tube and ovary or cushion-shaped and enveloping the base of the ovary. Stamens (5–)8–10(–20), rarely completely free, usually partly or completely fused to form a staminal tube, usually bearing appendages; anthers 2-thecous, dehiscing longitudinally, connective usually apiculate beyond anther-lobes. Ovary superior, (2–)4–5(–20)-locular, with axile placentation; ovules 1–many per locule; style 1; style-head expanded, capitate, globose, ovoid, cylindric, discoid or coroniform, entire or shallowly lobed, only partly stigmatic, sometimes (*Turraea*) functioning as a 'receptaculum pollinis'. Fruit usually a loculicidal or septifragal capsule or a drupe, rarely a leathery cleistocarp or a berry. Seeds usually arillate, winged or with a corky or woody outer covering; endosperm present or absent.

50 genera and about 800 species; almost confined to the tropics and subtropics of both hemispheres. 47 indigenous species in 14 genera in the Flora area, of which 6 species are endemic.

The following conspectus of the indigenous and naturalized genera is adapted from the generic monograph of Pennington & Styles (in Blumea 22: 419–540 (1975)) and uses only easily observed characters. It applies only to species occurring in East Africa.

Subfamily MELIOIDEAE. Buds naked. Fruit a loculicidal capsule, or a berry, drupe or cleistocarp. Seeds with a fleshy aril or exarillate, never winged.

Tribe **Turraeeae** *Harms*. Leaves simple. Indumentum of simple hairs. Flowers bisexual. Staminal tube complete, usually with appendages which are completely free or partly to completely united. Style-head forming a 'receptaculum pollinis'. Fruit a capsule with arillate seeds. Genus 1.

Tribe **Melieae**. Leaves pinnate or bipinnate. Indumentum of simple or stellate hairs. Flowers bisexual or male. Staminal tube complete, with appendages. Style head slightly expanded, shallowly lobed. Fruit a drupe. Genera 2, 3.

Tribe **Trichilieae** *DC*. Leaves pinnate. Indumentum of simple or stellate hairs. Flowers bisexual or unisexual. Staminal tube complete or incomplete, with or without appendages. Style-head slightly expanded, shallowly grooved or with indistinct apical stigmatic lobes. Fruit a capsule with arillate seeds, or a berry or a drupe. Genera 4–7.

Tribe **Guareeae** *Pennington & Styles*. Leaves pinnate. Indumentum of simple hairs. Flowers unisexual. Staminal tube complete, anthers inserted within the throat, appendages almost or completely fused. Style-head discoid. Fruit a capsule or cleistocarp. Seeds arillate. Genera 8, 9.

1

Subfamily SWIETENIOIDEAE Harms. Buds protected by scales. Fruit a septifragal capsule. Seeds winged or with a woody or corky aril.

Tribe **Cedreleae** *DC.* Leaves pinnate. Flowers with an androgynophore to which the petals are adnate by a carina. Stamens 5, free, without appendages. Capsule woody, columella well developed. Seeds winged. Genera 10, 11.

Tribe **Swietenieae** *A. Juss.* Leaves pinnate. Flowers without a gynophore to which the petals are adnate. Stamens 8 or 10. Staminal tube usually complete or almost so, with or without appendages. Capsule woody, columella well developed. Seeds winged. Genera 12–15.

Tribe **Xylocarpeae** *M. Roemer.* Leaves pinnate. Flowers without a gynophore to which the petals are adnate. Stamens 8–10(–12). Staminal tube complete, without appendages. Capsule leathery; columella poorly developed. Seeds unwinged, with a woody or corky aril. Genera 16, 17.

In many genera the flowers are unisexual but sterile vestiges of the non-functional sex are always well developed and, on a superficial examination, can easily be mistaken for fertile organs. In all cases the antherodes are indehiscent and lack pollen. The pistillodes may have vestigial locules and minute vestigial ovules, but they are always much more slender than the fertile gynoecia. In most of the older literature, the flowers of Meliaceae are described as hermaphrodite ('saepissime hermaphroditi', Bentham in Bentham & Hooker, G. P. 1: 327 (1862)). There are, however, many exceptions (Styles in Silvae Genetica 21: 175–182 (1972)). Except in *Turraeeae*, in most species the flowers are either male and female and borne *either* on different individuals (dioecious, as in *Ekebergia, Pseudobersama* and *Trichilia*) *or* in the same inflorescence (monoecious, as in subfamily *Swietenioideae*). In *Azadirachta* and *Melia* male and hermaphrodite flowers occur in the same inflorescence. Sex reversal has been reported for *Trichilia retusa* though possibly erroneously (see Styles, op. cit.: 181). In the female flowers of some species the disk appears to be less well developed than in the male; this should be looked into in the living plant in relation to floral mechanism as a whole.

The term 'aril' in this account refers to a red, yellow or white fleshy structure intimately associated with the seed; within the family its morphological nature varies greatly (see Harms in E & P. Pf., ed. 2, 19b(1): 25–26 (1940)) and in most cases is not known; for that reason we do not attempt to distinguish between aril, arillode and sarcotesta.

In *Guarea mayombensis* the fruit is indehiscent and has a leathery inedible pericarp surrounding large arillate seeds. Similar fruits are found in several Far Eastern species in other genera. Traditionally such fruits have been referred to as berries, but they are so different from the more familiar berries that we use a new term (for angiosperms), cleistocarp, here.

The Meliaceae, unlike some other tropical woody families, shows a considerable range (2n = 28 to 2n = c. 360) in chromosome number (Styles & Vosa in Taxon 20: 485–499 (1971); Khosla & Styles in Sylvae Genetica 24: 73–83 (1975)).

Chemically the Meliaceae is characterized by the presence of limonoids, a group of oxidized triterpenes. According to Taylor (in Progess in the Chemistry of Organic Natural Products 45: 1–102, (1984)) more than 250 limonoids have been extracted and identified from 54 species belonging to 23 genera; many other limonoids, some from additional species, have been subsequently discovered. Limonoids have been found to occur in nearly every member of the family so far investigated. Elsewhere they are of very restricted occurrence. In their limonoid chemistry the members of the Meliaceae form an extremely tightly knit and sharply defined group. Limonoids, e.g. azadirachtin, are powerful insect antifeedants and are used to control insect larvae including locusts (F. White in Bothalia 16: 155 (1986)).

In the text which follows, the keys and descriptions are relatively brief because many features are shown (more clearly than can be expressed in words) in the illustrations. In the species descriptions, atypical variation (e.g. very rarely occurring locule numbers) is not recorded unless its absence might cause confusion. Leaf measurements are based on fully expanded leaves on fertile shoots. Leaf length for pinnate leaves includes the petiole. Leaflet shape usually describes the ultimate or penultimate lateral leaflets. Juvenile and coppice leaves are not included in the descriptions. It should be noted, however, that the leaves of seedlings and young plants are sometimes strikingly different from those of the adult (see *Turraea wakefieldii*) but for most species this information is not available. In cymose inflorescences a clear distinction between bract and bracteole cannot be drawn and is not attempted here. Some of the illustrations of 'half flowers' are unconventional in that the main vertical section is displaced to show the whole of the style and style-head; a second section which shows placentation is more truly median.

Characters sometimes found elsewhere but not applicable in the Flora area: indumentum of peltate scales; leaves decussate, leaf-rhachis ending in a terminal 'bud'; inflorescence cauliflorous; flowers epiphyllous; ovary unilocular; fruit a nut.

The most recent revision of the Meliaceae at generic level is that of Pennington & Styles (in Blumea 22: 419–540 (1975)). It has stood the test of time and been widely used with little or no modification. In his Introduction, F. White (p. 422) remarked on the need for local floristic studies (such as this one) from which future syntheses would be derived. He also made a plea for the investigation of the development and differentiation of the flower, fruit and seed, especially the early stages, and for field studies of pollination and dispersal. Much remains to be done, but Pannell's work on the Far Eastern genus, *Aglaia* Lour. (D. Phil. thesis, Univ. of Oxford (1980); in Phil. Trans. R. Soc. Lond. B 316: 303–333 (1987)) could serve as a model for African genera. Developmental studies languish, though Cheek (D. Phil. thesis, Univ. of Oxford, (1990); in Mitt. Inst. Allg. Bot. Hamburg 23b: 683–706 (1990)) has described the structure of the mature seed of certain genera.

In both editions of Die Natürlichen Pflanzenfamilien, Harms included *Ptaeroxylon* Eckl. & Zeyh. in Meliaceae. Following Leroy (in Comptes Rendus Acad. Sc., Paris 248: 106 (1959) and Journ. Agric. Trop. Bot. Appl. 7: 456 (1960)), we place it in *Ptaeroxylaceae*.

Economically the Meliaceae is important, especially for its high-quality timbers, but there are countless other uses, both in East Africa and elsewhere. We have recorded only a few of them.

This work began in 1960 as part of a monograph of all the African species. It was supported in part by a Government grant (from the former Colonial Office). Long before the monograph was finished the nature of the employment of both its authors changed leaving little time for its completion. Hence the delay over the present contribution. The field studies on which this flora account is based were undertaken by Styles in Uganda and White in Kenya, Tanzania and some other African countries. During different phases of the work one or other author was more active than the other but we have both been involved in all its aspects and are jointly responsible for it.

Several exotic species are planted in East Africa for ornament or because of their actual or potential economic importance. They are briefly mentioned below and those that are locally naturalized or feature prominently in the landscape receive a fuller treatment in the main text.

Leaves 2–3-pinnate

1. *Melia azedarach* L. — see p. 22.

Leaves once pinnate; fruit a cleistocarp or drupe

2. *Aglaia odorata* Lour. Native of Indochina; widely planted in the tropics. Shrub or small tree up to 10 m. tall. Leaflets 3–5, glabrous, venation reticulate on both surfaces. Flowers very small, in axillary panicles, strongly scented, used (elsewhere) for flavouring tea and perfuming clothes. Fruit a small cleistocarp. Planted in Zanzibar (Victoria Gardens, *Dept. Agr.* 5!) and Kenya (Mombasa, *Parker* H76/35!).

3. *Azadirachta indica* A. Juss. — see p. 27.

4. *Lansium domesticum* Correa; Dale, Introd. Trees Uganda: 47 (1953); 'Langsat'; native of SE. Asia; sometimes planted elsewhere for its edible aril. Tree up to 15 m. tall. Leaflets 5–7. Flowers in spikes or panicles on older branchlets. Fruit a cleistocarp 4 cm. in diameter, yellow, with 1–3 seeds, each surrounded by a white, translucent, juicy, aromatic aril. Planted at Entebbe.

5. *Sandoricum indicum* Cav.; T.T.C.L.: 318 (1949); 'Santol'; native of tropical Asia. Much branched tree 10 m. tall (up to 20–30 m. in Asia). Leaflets 3–5, ovate-oblong, up to 20 × 12 cm., apex obtuse or shortly acuminate, base subtruncate; lower surface softly pubescent. Flowers in congested, axillary panicles. Fruit an orange, tomentose, leathery cleistocarp, ± 4 × 3.5 cm.; aril edible. Planted at Amani in 1902, *Greenway* 945!

Leaves once pinnate; fruit a woody capsule with winged seeds

6. *Cedrela odorata* L. — see p. 45.

7. *Swietenia macrophylla* King; Dale, Introd. Trees Uganda: 66 (1953); *Ekebergia sp. 4* of I.T.U.: 94 (1940); Honduras Mahogany; native of Central and South America. Tall tree. Leaflets 6–12, up to 15 × 5 cm., asymmetrically lanceolate, apex shortly acuminate. Flowers in axillary panicles. Capsule 12–15(–22) cm. long. Planted in Uganda for forestry purposes (without conspicuous success) and ornament. Entebbe, *Eggeling* 3818!

8. *Swietenia mahagoni* (L.) Jacq.; T.T.C.L.: 318 (1949); Dale, Introd. Trees Uganda: 66 (1953); 'Spanish or Cuban Mahogany', native of the West Indies and Florida. Tall tree with smaller, more attenuate leaflets and smaller fruits than *S. macrophylla*. Occasionally planted in Uganda, Kenya (e.g. Mwachi Forest, *Faden et al.* 77/450!) and Tanzania (e.g. Dar es Salaam, *Wigg* 958!).

9. *Toona ciliata* M. Roemer — see p. 45.

10. *T. serrata* (Royle) M. Roemer — see p. 46.

Leaves imparipinnate, pinnately 3-foliolate, 1-foliolate, simple
    or 2–3-pinnate; fruit a berry or a drupe or a small
    loculicidal capsule; seeds neither winged nor with a corky
    covering, often arillate:
Leaves simple; style-head expanded to form a 'receptaculum
    pollinis'; fruit a small capsule; seeds arillate . . .   **1. Turraea**
Leaves compound:
    Leaves 2–3-pinnate or pinnate; staminal tube complete,
      with appendages; fruit a drupe:
      Leaves 2–3-pinnate; indumentum of simple and stellate
        hairs . . . .   **2. Melia**
      Leaves pinnate; indumentum of simple hairs . . .   **3. Azadirachta**

Leaves pinnate; if fruit drupaceous then staminal tube incomplete (*Lepidotrichilia*) or without appendages (*Ekebergia*):

  Anthers inserted apically on the margin of the staminal tube or filaments; petals not fused to the staminal tube for the greater part of its length:

    Fruit a capsule:

      Capsule with 5 thick woody valves covered with conspicuous ridges and antler-like appendages   **4. Pseudobersama**

      Capsule with 2–3(–4) leathery valves without antler-like appendages, but occasionally (*T. lovettii*) verruculose . . . .   **5. Trichilia**

    Fruit a drupe:

      Indumentum stellate; staminal tube incomplete, filaments with paired appendages   . . .   **6. Lepidotrichilia**

      Indumentum simple; staminal tube complete, without appendages . . . .   **7. Ekebergia**

  Anthers inserted within the throat of the staminal tube:

    Petals fused to the staminal tube for the greater part of its length; petioles not grooved and winged at base . . . .   **8. Turraeanthus**

    Petals free from staminal tube; petioles deeply grooved and winged at base   . . . . .   **9. Guarea**

Leaves paripinnate; fruit a septifragal capsule; seeds winged or with a thick corky or woody outer covering:

  Flowers with an androgynophore to which the bases of the petals are adnate; stamens 5, free:

    Androgynophore much longer than the ovary; first leaflets of seedlings entire . . . .   **10. Cedrela**

    Androgynophore shorter than or equalling the ovary; first leaflets of seedlings lobed or toothed   . . . .   **11. Toona**

  Flowers without an androgynophore; petals free; stamens 8–10, filaments partly or completely united:

    Capsule with a well-developed columella; seeds winged:

      Capsule globose or subglobose; seeds orbicular to suborbicular, winged all the way round   . . .   **12. Khaya**

      Capsule elongate, at least twice as long as broad; seeds with a terminal wing:

        Seeds attached by the seed-end towards the distal end (apex) of the columella (i.e. seed winged below); flowers 5-merous:

          Leaflets entire; capsule pendulous; valves lacking a fibrous network; staminal tube lacking appendages . . . .   **13. Entandrophragma**

          Leaflets undulately lobed; capsule erect; valves held together by a fibrous network; staminal tube with deltate appendages   . . . . .   **14. Pseudocedrela**

        Seeds attached by the wing-end towards the distal end (apex) of the columella (i.e. seed winged above); flowers 4-merous   . . . .   **15. Lovoa**

    Capsule with a rudimentary columella; seeds unwinged, with a corky or woody outer covering:

      Inflorescence much branched; sepals imbricate, ± free to base; leaflets in 6–18 or more pairs; seeds with a woody covering   . . . .   **16. Carapa**

      Inflorescence little branched; sepals valvate, united in the lower half; leaflets in (1–)2–3(–4) pairs; seeds with a corky covering   . . . .   **17. Xylocarpus**

1. TURRAEA

L., Mant. Pl. Alt.: 150, 237 (1771); Pennington & Styles in Blumea 22: 455 (1975)

Shrubs or small trees, sometimes scrambling. Leaves simple, alternate. sometimes fasciculate, usually entire, rarely repand, undulate or shallowly lobed. Flowers bisexual, solitary or fasciculate or in axillary or terminal cymes or false racemes. Calyx cupuliform, with (4–)5(–6) teeth or lobes, sometimes almost entire margined, persistent in fruit. Petals (4–)5, much longer than the calyx, linear-spathulate to linear, imbricate or contorted. Staminal tube very variable (see below), cylindric, sometimes expanded distally, terminated by simple or 2-lobed, free or partly or completely fused appendages opposite to or alternating with the anthers, or appendages rarely absent; anthers (8–)10, apiculate, inserted on the rim of the staminal tube or inside the tube towards the apex, sessile or with filaments. Disk usually present, sometimes vestigial, partly or completely fused to the base of the staminal tube. Ovary small, with (4–)5–12 locules, each with 2 anatropous or almost campylotropous superposed ovules; style elongate, exserted, distally expanded to form an ovoid, globose, cylindric or conical style-head of which the proximal part serves as a 'receptaculum pollinis' and the distal part forms a disk- or cushion-shaped stigmatic surface. Fruit a small leathery or woody loculicidal capsule with (4–)5–10 or 12 valves. Seeds reniform with a large or small, usually conspicuous, white, orange or red aril; testa black or red, shining.

About 50 species in Africa, Madagascar, the Mascarenes and Comores plus a few in the tropical Far East.

Characters not applicable in the Flora area: leaves rarely compound; calyx-lobes rarely foliaceous; style-head rarely included, and then only slightly expanded and coroniform (see Bothalia 16, fig. 4 (1986)).

There is a great variation in the structure of the staminal tube, especially in its appendages. This is presumably related to pollination, though precise observations, apparently, have not yet been made.

In *T. pulchella* (Harms) Pennington, a South African species (Bothalia 16, fig. 4/1a (1986)), the filaments are free for a short distance at the apex, and each bears a pair of filiform appendages inserted one on either side at the base of the anthers. In most other African species the filaments are completely fused and the appendages occur in pairs on the rim of the tube alternating with the anthers.

In *T. floribunda, T. kokwaroana, T. mombassana* and *T. wakefieldii* the members of a pair are free to the base or almost so; in *T. mombassana* (fig. 2/2, p. 15) they are laciniate. In *T. barbata* (fig. 1/2, p. 9), *T. elephantina, T. fischeri* (fig. 1/4), *T. parvifolia* (fig. 1/6) and *T. vogelii* (fig. 1/3) the filaments are fused in pairs at least in their lower half, but each pair remains free from the others.

In the remaining East African species all the appendages are fused (at least at the base) to form a staminal frill. The fusion is incomplete in *T. holstii* (fig. 1/5), *T. nilotica* (fig. 1/1) and *T. robusta*. Fusion is normally complete or almost so in *T. abyssinica, T. cornucopia* (fig. 1/6), *T. kimbozensis, T. pellegriniana* and *T. stolzii*.

In *T. vogelioides* (fig. 3, p. 17) appendages are normally absent.

The fruits of *Turraea* are also very variable within the genus and provide useful features for classification, though many will remain elusive until they can be described from living material. There is a broad correlation between groups of species based on floral characters and those based on fruits, but some species are discordant.

Most African species of *Turraea* are well defined though some species are very variable, sometimes in several features. This makes for difficulties in identification and care should be excercised. The keys provided below are meant as simple aids. They are not meant to be foolproof. They should be used in conjunction with the descriptions and, especially, the illustrations.

KEY BASED ON FLORAL AND VEGETATIVE CHARACTERS*

Filaments well developed, geniculate, usually at least half as
  long as anthers; staminal tube bearded at the throat, the
  hairs arising at least in part from the filaments and visible
  from outside without dissection:
Staminal appendages free to the base, bifid for more than
  half of length; filaments about half as long as anthers   1. *T. barbata*
Staminal appendages fused in lower half or to beyond the
  middle to form a frill continuing staminal tube beyond
  insertion of filaments; filaments about as long as
  anthers:

*See also key based on fruit and vegetative characters, p. 7.

Filaments hairy for whole of length; ovary densely pilose
(exceptionally glabrous); peduncles (0.5–)0.7–4.5 cm.
long; bracts foliaceous, the longest usually more than
0.5 cm. long; outside of petals usually coarsely
tomentellous; flowers produced with the leaves   2. *T. robusta*
Filaments hairy only at the base; ovary glabrous;
peduncles absent or rarely up to 0.5 cm. long; bracts
subulate, up to 0.3 cm. long; outside of petals sparsely
puberulous to finely tomentellous; usually flowering
when leafless . . . .   3. *T. nilotica*
Filaments vestigial; staminal tube glabrous or variously hairy
inside but never bearded at the throat; hairs not visible
without dissection:
Hairs present in upper half of staminal tube on the inside:
Staminal appendages fused for greater part of length to
form a frill continuing the staminal tube beyond the
insertion of the anthers:
Leaf-margin repand or sinuate; inflorescence a 7–15-
flowered axillary, congested cyme or false raceme;
bracts 0.1 cm. long   . . . .   4. *T. kimbozensis*
Leaf-margin entire:
Bracts foliaceous, 0.3–0.8 cm. long; inflorescence a
4–12-flowered cyme or false raceme; calyx lobed
almost to the middle, tomentellous   . . . .   5. *T. abyssinica*
Bracts squamiform, 0.1–0.2 cm. long; calyx
puberulous:
Inflorescence a subsessile fascicle with up to 19
flowers . . . .   6. *T. pellegriniana*
Inflorescence a (1–)2–3(–4)-flowered, distinctly
pedunculate cyme:
Staminal tube slightly expanded at the throat;
anthers partly visible from outside; pedicels
(0.7–)2–4.3 cm. long, usually very slender;
calyx 0.2–0.3 cm. long   . . . .   7. *T. holstii*
Staminal tube markedly expanded at the throat;
anthers completely concealed from outside;
pedicels 1.3–2 cm. long, stout; calyx 0.4–0.6
cm. long . . . .   8. *T. stolzii*
Staminal appendages not fused to form a frill:
Leaves minute, up to 4.2 × 0.8 cm., in fascicles with the
flowers on short spur shoots; staminal appendages
well developed . . . .   9. *T. parvifolia*
Leaves much larger, separated by long internodes;
flowers in distinctly pedunculate inflorescences:
Climbing shrub; leaves unlobed; staminal
appendages well developed   . . . . . .   10. *T. vogelii*
Erect shrub; leaves slightly repand to shallowly lobed;
staminal appendages absent or vestigial; anthers
borne on rim of staminal tube   . . . . .   11. *T. vogelioides*
Hairs absent from upper half of staminal tube:
Staminal tube curved, trumpet-shaped, gradually narrowed
from the apex to the base:
Staminal appendages ± completely fused to form a frill
beyond the insertion of the anthers, which are
completely included   . . . .   12. *T. cornucopia*
Staminal appendages not fused to form a frill; anthers
exserted:
Staminal tube 4 cm. long; style-head conspicuously
rostrate   . . . .   13. *T. elephantina*
Staminal tube 1.6–2.4 cm. long; style-head ovoid, not
rostrate:

Leaves broadest just above the base, setulose
 beneath; apex rounded to subacute, base
 cordate or subtruncate; staminal appendages
 ± equalling the anthers, free to the base 14. *T. kokwaroana*
Leaves broadest well above the base with longish
 hairs or glabrous beneath; apex acute to
 shortly acuminate; base cuneate or rounded
 (and then mostly shortly decurrent); staminal
 appendages shorter than the anthers, partly
 fused in pairs . . . . 15. *T. fischeri*
Staminal tube straight, narrowly cylindric throughout or
 suddenly expanded at the apex; staminal appendages
 filiform, much longer than the anthers:
Style-head conoidal-cylindric, tapering from a broad
 base; leaves with conspicuous tufts of hairs in
 nerve-axils beneath:
Flowers solitary or fasciculate among the leaves;
 ovary glabrous, 5(–6)-locular . . . . . . 16. *T. mombassana*
Flowers in leaf-axils or in axils of fallen leaves; ovary
 tomentellous, 10(–12)-locular . . . . . . 17. *T. wakefieldii*
Style-head globose; leaves without conspicuous tufts of
 hairs; ovary ± glabrous, 10(–12)-locular . . . 18. *T. floribunda*

KEY BASED ON FRUIT AND VEGETATIVE CHARACTERS

NOTE. The only available fruits of species 1. *T. barbata* are immature; fruits are unknown for 13. *T. elephantina*.

Leaves, flowers and fruits mostly borne on short lateral shoots
 of slow and limited growth:
Seeds red, aril white . . . . 16. *T. mombassana*
Seeds black, aril red or orange:
Leaf-lamina up to 4.2 × 0.8 cm. . . . . 9. *T. parvifolia*
Leaf-lamina longer and much broader:
Leaf-lamina (at least of some leaves) subcordate at the
 base:
Leaf-lamina mostly broadest near the middle,
 tapering gradually towards the apex . . . . 1. *T. barbata*
Leaf-lamina broadest near the base, tapering rapidly
 towards the apex . . . . 14. *T. kokwaroana*
Leaf-lamina cuneate to broadly rounded at the base:
Leaf-lamina tapering rapidly from near the base to
 the acute to subacuminate apex, base usually
 cuneate . . . . 15. *T. fischeri*
Leaf-lamina tapering gradually from near the middle
 to the obtuse, rounded or slightly emarginate
 apex, base not or very rarely cuneate:
Tertiary nerves and veins conspicuous on lower
 leaf-surface . . . . 13. *T. elephantina*
Tertiary nerves and veins inconspicuous . . . 12. *T. cornucopia*
Leaves mostly borne on long shoots, usually widely spaced:
Capsule nearly always longer than broad, very rarely
 depressed-globose but then valves very thick and woody
 and aril white:
Septa of capsule conspicuous, vermilion, membranous;
 seeds black, aril vermilion:
Climbing shrub; leaves unlobed . . . . 10. *T. vogelii*
Erect shrub; leaves slightly repand to shallowly lobed 11. *T. vogelioides*
Septa not as above; seeds orange, aril white:
Capsule glabrous, valves very thick (± 0.5 cm.) and
 woody . . . . 18. *T. floribunda*
Capsule tomentellous; valves thinner . . . . . . 17. *T. wakefieldii*

Capsule depressed-globose, valves not thick and woody,
  septa not vermilion; seeds dark red or black, aril orange
  or red:
Leaf-lamina distinctly acuminate or subacuminate:
  Leaf-margin repand or sinuate . . . .   4. *T. kimbozensis*
  Leaf-margin entire:
    Infructescence a distinctly pedunculate cyme:
      Fruiting pedicels conspicuously wider at apex
        below the fruiting calyx; lower leaf-surface
        with conspicuous venation and inconspicuous
        axillary tufts of hairs . . . .   8. *T. stolzii*
      Fruiting pedicels not conspicuously wider at apex;
        lower leaf-surface with conspicuous axillary
        tufts of hairs:
          Tertiary nerves and veins conspicuous on lower
            leaf-surface, fruiting pedicels usually less
            than 1.5 cm. long . . . .   5. *T. abyssinica*
          Tertiary nerves and veins inconspicuous on
            lower leaf-surface; fruiting pedicels usually
            more than 2 cm. long. . . . . . .   7. *T. holstii*
    Infructescence a subsessile fascicle . . . . .   6. *T. pellegriniana*
Leaf-lamina rounded or obtuse, rarely acute or apiculate:
  Bark on older branchlets often corky; inflorescence a
    sessile or subsessile fascicle. . . . . . .   3. *T. nilotica*
  Bark not corky; infructescence pedunculate . . .   2. *T. robusta*

1. **T. barbata** *Styles & F. White* in B.J.B.B. 59: 257 (1989). Type: Kenya, Northern Frontier Province, Wajir, *Dale* K717 (EA, holo.!, K, iso.!)

Shrub or small tree up to 4 m. tall; lateral shoots slow growing, with crowded internodes. Leaf-lamina ovate to ovate-elliptic, up to 6 × 4.5 cm., generally puberulous with short stiff appressed setulose hairs on both surfaces and especially on the nerves beneath, apex rounded or emarginate, base broadly rounded to subcordate; petiole up to 0.8 cm. long. Inflorescence an axillary or terminal congested false raceme; peduncle short, up to 0.4 cm. long; bracts subulate, up to 0.2 cm. long, puberulous; pedicels up to 1.3 cm. long. Flowers white or creamy-white. Calyx ± 0.3 cm. long, tomentellous, lobed to ± one-third. Petals linear to linear-spathulate, 1.5–1.9 cm × 0.2–0.3 cm., tomentellous outside. Staminal tube 0.7–1.2 cm. long, expanded at the apex, bearded at the throat with long hairs arising from the filaments and from the uppermost part of the staminal tube, otherwise glabrous inside, with a few scattered hairs outside; appendages deeply 2-fid for more than half of length, alternating with anthers; not fused at the base, ± 0.35 cm. long, glabrous outside. Ovary 10–11-locular, tomentellous; style 1.4–1.7 cm. long, sometimes puberulous at the base, otherwise glabrous; style-head globose-cylindric, its base exserted ± 0.7 cm. beyond staminal-tube. Capsule (immature) depressed-globose, ± 0.6 × 0.8 cm., tomentellous. Fig. 1/2.

KENYA. Northern Frontier Province: Dandu, Mar. 1952, *Gillett* 12526! & 40 km. Wajir–Tarbaj, May 1972, *Gillett* 19757! & Dadaab–Wajir, 1 km. N. of Uaso Nyiro R. crossing, May 1977, *Gillett* 21237!
DISTR. **K** 1, 7; Somalia
HAB. In *Acacia-Commiphora* deciduous bushland; 70–610 m.

SYN. *T. sp. 1* sensu Dale & Greenway, K.T.S.: 275 (1961)
     T. sp. 2 sensu Dale & Greenway, K.T.S.: 276 (1961), pro parte quoad *Gillett* 12526 saltem

NOTE. The leaves of *T. barbata* are similar to those of *T. kokwaroana* in shape and indumentum, but its flowers are very different.

2. **T. robusta** *Gürke* in E.J. 19, Beibl. 47: 34 (1894) & in P.O.A. C: 231 (1895); Bak.f. in J.B. 41: 12 (1903); Z.A.E.: 433 (1912); Siebenlist, Forstwirtschaft in Deutsch-Ostafrika: 92 (1914); V.E. 3(1): 815 (1915); T.T.C.L.: 321 (1949); I.T.U., ed. 2: 200 (1952); K.T.S.: 275 (1961); F. White & Styles in F.Z. 2: 308, t. 61C (1963); Styles in E. Afr. Agric. Journ. 39: 421 (1974). Type: Tanzania, Usambara Mts., near Gonja, *Holst* 9069 (B, holo.†, K, iso.!)

Shrub or small to medium-sized tree (1–)2–8(–16) m. tall, sometimes weak-stemmed and scrambling; bark light grey or brown, slightly fissured. Leaf-lamina elliptic, obovate or obovate-elliptic, usually less than 11 × 7 cm., lower surface usually shortly pubescent

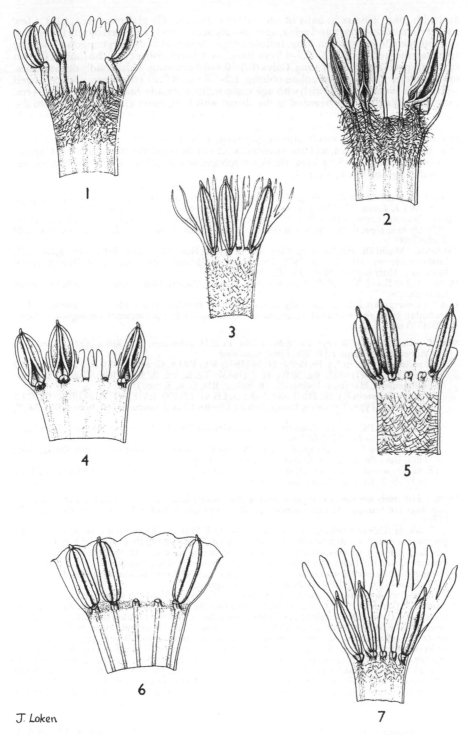

FIG. 1. *TURRAEA SPP.*—details of staminal tube, × 6. **1**. *T. NILOTICA;* **2**, *T. BARBATA;* **3**, *T. VOGELII;* **4**, *T. FISCHERI;* **5**; *T. HOLSTII;* **6**, *T. CORNUCOPIA;* **7**, *T. PARVIFOLIA.* 1, from *Bullock* 3202; 2, from *Dale* 717; 3, from *Keay* s.n.; 4, from *B.D. Burtt* 5129; 5, from *Greenway* 752; 6, from *Archer* 123; 7, from *Bally* 12616. Drawn by Julia Loken.

and with tufts of hairs in axils of secondary nerves, rarely glabrous except for a few scattered hairs on nerves and veins, apex usually acute, rarely rounded or apiculate, base cuneate; petiole up to 1.8 cm. long. Inflorescence a terminal or axillary congested cyme or false raceme; peduncle (0.5–)0.7–4.5 cm. long, stout; bracts foliaceous, the longest 0.5–1.5 cm. long; pedicels 0.8–2 cm. long. Calyx (0.3–)0.4–0.6 cm. long, with broadly deltate lobes, tomentellous. Petals oblanceolate-oblong, 1.2–1.8 cm. × 0.45 cm., creamy white tinged with green, becoming yellowish with age, tomentellous outside. Staminal tube 0.8–1.4 cm. long, distally expanded, bearded at the throat with long hairs arising mostly from the filaments and a few from the appendages, otherwise glabrous inside; appendages 1–3-lobed, fused for about half of length, narrowly deltate at apex. Ovary 10–12(–14)-locular, densely pilose or very rarely almost glabrous; style 1.5–2.2 cm. long, pilose at base; style-head broadly ovoid, its base exserted ± 0.2 cm. beyond the staminal tube. Capsule depressed-globose, ± 0.8 × 1.5 cm., shallowly sulcate, woody, pilose or very rarely glabrous. Seeds dark red or black; aril red.

UGANDA. Kigezi District: Bwambaga [Buambara], Feb. 1950, *Purseglove* 3274!; Busoga District: Jinja, Aug. 1904, *E. Brown* 83!; Mengo District: Mabira Forest, Nov. 1919, *Dummer* 4339!
KENYA. Nairobi District: 24 km. Nairobi–Thika, Feb. 1951, *Greenway & Verdcourt* 8492!; Kisumu, Nov. 1939, *Opiko in Herb. Bally* 666!; Masai District: Migori Bridge, June 1961, *Glover, Gwynne, Samuel & Tucker* 1806!
TANZANIA. Moshi District: Njoro R., 8 km. west of Moshi, Nov. 1955, *Milne-Redhead & Taylor* 7045!; Lushoto District: Mkuzi, Apr. 1953, *Drummond & Hemsley* 2080!; Morogoro District: above Morogoro, Morningside, Mar. 1951, *Eggeling* 6064!
DISTR. U 2–4; K 4–7; T 1–3, 5–8; Zaire, Burundi, Zambia, Malawi, Mozambique; sometimes planted for ornament
HAB. In evergreen forest, especially at the edges of riparian and mid-altitude forest, and in secondary scrub-forest or as a forest remnant; sometimes in fire-protected *Brachystegia* woodland; 900–1750 m.

SYN. *T. nilotica* Kotschy & Peyr. var. *robusta* Oliv. in H.H. Johnston, The Kilimanjaro Expedition, Appendix to Chapter 17: 339 (1886), *nom. nud.*
    T. volkensii Gürke in E.J. 19, Beibl. 47: 34 (1894) & in P.O.A. C: 231 (1895); Bak.f. in J.B. 41: 12 (1903); V.E. 3(1): 815, fig. 384G, H, J (1915); T.S.K., ed. 2: 104 (1936). Type: Tanzania, Kilimanjaro, Marangu, *Volkens* 257 (B, holo.†, BM, G, K, Z, iso.!)
    T. goetzei Harms in E.J. 28: 415 (1900); Bak.f. in J.B. 41: 12 (1903); V.E. 3(1): 815 (1915); T.T.C.L.: 321 (1949). Type: Tanzania, Iringa District, Uhehe, Ufuagi, *Goetze* 747 (B, holo.†, BM, BR, P, iso.!)
    T. sacleuxii C. DC. in Ann. Conserv. Jard. Bot. Genève 10: 130 (1907). Type: Kenya, Teita Hills, Bura Peak, *Sacleux* 2355 (P, holo.!)
    T. squamulifera C. DC. in Ann. Conserv. Jard. Bot. Genève 10: 133 (1907). Type: Kenya, Teita Hills, Bura Peak, *Sacleux* 2355 (P, holo.!) — see note
    [*T. nilotica* sensu Staner in B.J.B.B. 16: 127, fig. 3 (1941); F.P.N.A. 1: 419 (1948); Staner & Gilbert in F.C.B. 7: 156 (1958), *non* Kotschy & Peyr.]

NOTE. The fruits are said to be poisonous (*Koritschoner* 1468). In Moshi District, Tanzania, wild seedlings are transplanted into banana gardens and grown as fodder for goats (*Wigg* in F.H. 1431).
    T. robusta is closely related to *T. nilotica* Kotschy & Peyr., and has been confused with it. Its geographic range lies almost entirely within that of *T. nilotica* but their ecology is different. *T. robusta* is a forest species whereas *T. nilotica* occurs in savanna. All specimens of *T. robusta* examined from the Flora area differ from *T. nilotica* in *most* of the characters given in the key. A few specimens however which are otherwise typical have glabrous or sparsely pilose fruits (*Bruce* 1119, *Semsei* 1192, *Wigg* 1032, all from Morogoro District).
    In the Paris herbarium there are four sheets of *Sacleux* 2355 which have been annotated by Casimir de Candolle. One sheet, to which is attached the manuscript notes of the collector, is the holotype of *Turraea squamulifera* C. DC. Three other sheets with this number but lacking collector's notes have been labelled by de Candolle as "*T. sacleuxii* C. DC." which was described in the same paper as *T. squamulifera*. Sacleux sometimes gave the same number to different gatherings made at different times which he thought belonged to the same species. The manuscript labels usually provide evidence of this, but in the present case there is nothing to suggest that the types of *T. sacleuxii* and *T. squamulifera* are not parts of the same gathering. Three gatherings from outside the Flora area (*Lawton* 425 and 427 from Nkamba Bay, Lake Tanganyika, Zambia and *Bredo* 3169 from Kalangila, Fungurume, Shaba) are variously intermediate between *T. robusta* and *T. nilotica* and may be of hybrid origin.

3. **T. nilotica** *Kotschy & Peyr.*, Pl. Tinn.: 12, t. 6 (1867); F.T.A. 1: 331 (1868); C. DC. in A. & C. DC., Monogr. Phan. 1: 445 (1878); Gürke in P.O.A. C: 231 (1895); Bak.f. in J.B. 41: 11 (1903); V.E. 3(1): 815, fig. 384 L, M (1915); T.T.C.L.: 321 (1949); K.T.S.: 275 (1961); F. White & Styles in F.Z. 2: 310, t. 61D (1963) & in F.S.A. 18: 43, fig. 11/1 (1986); F. White in Bothalia

16: 149 (1986); F. White & Styles in Fl. Eth. 3: 481, fig. 124.2/4–6 (1990). Types: Sudan, near Req, *Heuglin* 54, near Bongo, *Heuglin* 55, near Wau, *Heuglin* 56 & near Gondokoro, *Knoblecher* (all W, syn.!)

Shrub or small deciduous tree up to 10 m. tall, occasionally flowering as a shrublet; bark rough, brown, flaking in squarish pieces; older branchlets stout, often with a thick corky bark. Leaf-lamina elliptic to oblanceolate or obovate, rarely lanceolate, usually less than 15 × 8 cm., lower surface densely and softly pubescent to glabrous, apex rounded or obtuse, rarely emarginate or apiculate, base cuneate; petiole up to 1.8 cm. long. Inflorescence a 5–18-flowered sessile or subsessile fascicle in the axils of fallen leaves and then often terminating short shoots of slow growth, exceptionally in leaf-axils; bracts subulate, up to 0.3 cm. long; pedicels 0.5–1.6 cm. long. Flowers fragrant, greenish white turning yellow with age, usually appearing before the leaves. Calyx ± 0.25 cm. long, denticulate, puberulous. Petals oblanceolate-oblong, 1.25–2.2 cm. × 0.3–0.45 cm., sparsely puberulous to (rarely) minutely tomentellous. Staminal tube 0.8–1.5 cm. long, distally expanded, bearded at the throat with long hairs arising from the lower half of the filaments and uppermost 0.2 cm. of staminal-tube, otherwise glabrous inside; appendages 0.15–0.25 cm. long, usually fused to beyond the middle to form a frill with a shallowly lobed subtruncate apex, glabrous outside or rarely with a few marginal cilia. Ovary with (8–)10(–12) locules, glabrous; style 1.6–2.5 cm. long, glabrous or pilose at the base; style-head ovoid-cylindric, its base exserted ± 0.8 cm. beyond the staminal tube. Capsule depressed-globose, ± 0.7 × 1.5 cm., shallowly sulcate, thinly woody, glabrous or rarely puberulous. Seeds 0.5 × 0.3 cm., black; aril orange or red. Fig. 1/1, p. 9.

KENYA. Kilifi District: Mazeras market, Feb. 1937, *Dale* in *F.D.* 3630!; Lamu Town, Feb. 1956, *Greenway & Rawlins* 8905!
TANZANIA. Pangani District: Bushiri Estate, Oct. 1950, *Faulkner* 724!; Kondoa District: Kikori, Aug. 1929, *B.D. Burtt* 2171!; Iringa District: 90 km. N. of Iringa, July 1956, *Milne-Redhead & Taylor* 11229!; Zanzibar I., Chwaka, Nov. 1959, *Faulkner* 2395!
DISTR. **K** 1, 4, 7; **T** 1–8; **Z**; Sudan, Ethiopia and Somalia, southwards to South Africa (Transvaal)
HAB. In woodland, bushland and wooded grassland; 0–1525 m.

SYN. *T. randii* Bak. f. in J.B. 37: 427 (1899) & 41: 11 (1903). Type: Zimbabwe, Harare, *Rand* 562 (BM, holo.!)
　　T. tubulifera C. DC. in Ann. Conserv. Jard. Bot. Genève 10: 133 (1907). Type: Tanzania, Bagamoyo, *Sacleux* 471 (P, holo.!)
　　T. nilotica Kotschy & Peyr. var *glabrata* Fiori in Chiov., Result. Sci. Miss. Stef.-Paoli, Coll. Bot. 1: 49 (1916). Types: Somalia, Sahenen–Biobahal, near R. Juba [Guiba], *Paoli* 849 & *Paoli* 981 (FT, syn.!)

VAR. *T. nilotica* is very variable in leaf-shape and indumentum and some of this variation shows significant geographical correlation, which is, however, not sufficiently definite to warrant the recognition of subspecies. The leaves of specimens from the coastal strip of Kenya and northern Tanzania (including Zanzibar) are usually glabrate on the lower surface when mature and are usually subacuminate to shortly acuminate at the apex, but are sometimes rounded to acute. The leaves of specimens from the inland parts of Tanzania and all countries to the south are characteristically tomentose or densely pubescent and rounded or subacute (*T. randii*), but sparsely pubescent leaves and shortly acuminate apices occur scattered through most of this part of the range. If this accounted for the whole of the variation of the species it might be possible to recognize two subspecies, but leaf-shape and indumentum are very variable in the outlying populations in the Sudan (which includes the types of *T. nilotica*) and at the north-eastern extremity of the species-range in Somalia (which includes the type of *T. nilotica* var. *glabrata*).

NOTE. Powdered roots are used as a purgative but have been known to cause death (Superintendent of Police, Tanzania, H51/49). Green leaves are said to be edible to cattle, but when dried are poisonous to man and beast (*Greenway* 801). Eggeling & Dale, I.T.U., ed. 2: 199 (1952) record *T. nilotica* from Uganda. It possibly occurs there but we have seen no specimens.

4. **T. kimbozensis** *Cheek* in K.B. 44: 465, fig. 1 (1989). Type: Tanzania, Morogoro District, Kimboza Forest Reserve, *Bidgood, Mwasumbi & Vollesen* 1240, sheet 1 (K, lecto. here designated! — see note)

Treelet 1–2 m. tall. Leaf-lamina elliptic or oblanceolate-elliptic, up to 21 × 10 cm., glabrous beneath except for inconspicuous tufts of hairs in the nerve-axils and a few scattered hairs on the main nerves, apex acuminate, base cuneate; margin slightly to distinctly repand (exceptionally deeply sinuate); petiole up to 0.5 cm. long. Inflorescence an axillary congested cyme or false raceme with 7–15 flowers; peduncle 0.2–0.4 cm. long; bracts ± 0.1 cm. long; pedicels 1–1.2 cm. long. Calyx ± 0.2 cm. long, denticulate, minutely puberulous outside. Petals ± 1 × 0.25 cm., minutely puberulous outside. Staminal tube ± 0.9 cm.

long, expanded distally, glabrous inside except for a ring of hairs 0.3 cm. below the insertion of the anthers; appendages fused to form a subentire frill completely concealing the anthers. Ovary 5–6-locular, glabrous. Style 1.2 cm. long; style-head globose, exserted 0.3–0.4 cm. beyond apex of staminal tube. Capsule depressed-globose, ± 1.5 × 1.8 cm., shallowly sulcate, thinly woody, glabrous. Seeds ± 0.8 × 0.45 cm., black; aril red.

TANZANIA.   Morogoro District; Kimboza Forest Reserve, June 1983, *Polhill & Lovett* 4906!
DISTR.   **T** 6; not known elsewhere
HAB.   Lowland (groundwater) rain-forest on limestone; 300–450 m.

NOTE.   Two sheets with the same number have been labelled by the author as holotypes. One has flowers, the other fruits. We designate the former as a lectotype.

5.   **T. abyssinica** *A. Rich.*, Tent. Fl. Abyss. 1: 106, fig. 25 (1847); F.T.A. 1: 331 (1868); C. DC. in A. & C. DC., Monogr. Phan. 1: fig. 61/5 (1878); Bak.f. in J.B. 41: 10 (1903); V.E. 3(1): 813, fig. 384F (1915); T.S.K., ed. 2: 103 (1936); K.T.S.: 272 (1961); Styles in E. Afr. Agric. Journ. 39: 420 (1974); Styles & F. White in Fl. Eth. 3: 481, fig. 124.2/7, 8 (1990). Types: Ethiopia, Tigray, Scholoda Mt., *Schimper* 28 (P syn.!, G, K, L, M, Z, isosyn.!) & *Schimper* 191 (P, syn.!, BR, FT, G, K, L, M, isosyn.!)

Shrub or small to medium-sized tree up to 12(–18) m. tall, sometimes scrambling. Leaf-lamina lanceolate or lanceolate-elliptic, up to 12 × 5 cm., glabrous except for conspicuous tufts of hairs in the nerve-axils and scattered hairs on the nerves beneath, apex narrowly and acutely acuminate, base cuneate, slightly unequal; petiole up to 1 cm. long. Inflorescence a terminal or lateral congested cyme, or false raceme, usually with 4–12 flowers; peduncle 0.7–2.5 cm. long; bracts foliaceous, 0.3–0.8 cm. long, tomentellous; pedicels 0.5–1.6 cm. long. Calyx ± 0.4 cm. long, lobed almost to middle, tomentellous. Petals linear-spathulate, 1.4–2.2 × 0.3–0.4 cm., greenish white or cream, densely puberulous to tomentellous outside. Staminal tube 1–1.4 cm. long, very narrowly cylindric, slightly expanded at the throat, glabrous inside immediately below the insertion of the anthers but with longish dense hairs in the remainder of the upper half inside, sometimes with scattered hairs on the outside; appendages fused for the greater part of their length to form an irregularly lobed frill 0.15–0.3 cm. long beyond the insertion of the filaments, almost completely concealing the anthers, glabrous outside. Ovary 5–6(–7)-locular, glabrous; style 1.4–1.8 cm. long, glabrous; style-head globose, large. Capsule depressed-globose, ± 0.7 × 0.8 cm., shallowly sulcate, thinly woody, glabrous. Seeds black; aril red.

UGANDA.   Karamoja District: Napak, June 1950, *Eggeling* 5947!
KENYA.   Northern Frontier Province: Mt. Nyiru, June 1936, *Lady Muriel Jex-Blake* 9!; Kiambu District: near Mt. Margaret, Feb. 1953, *Drummond & Hemsley* 1211!; Masai District: Ngong Hills, May 1960, *Polhill* 217!
TANZANIA.   Mbulu District: Basoda, Dec. 1927, *B.D. Burtt* 1701!
DISTR.   **U** 1; **K** 1–4, 6; **T** 2; Ethiopia
HAB.   In montane forest, usually of drier types, and at its edges; sometimes in upland riparian forest; persisting as a forest relic in montane grassland; 1820–2225 m.

SYN.   *T. kilimandscharica* Gürke in P.O.A. C: 230 (1895); Bak.f. in J.B. 41: 10 (1903); V.E. 3(1): 813 (1915); T.T.C.L.: 321 (1949). Type: Tanzania, Kilimanjaro, Useri, *Volkens* 2004 (B, holo.†, BM, G, Z, iso.!)

NOTE.   The wood is easy to work and is used for making huts and furniture. The Masai use the roots in medicine (*Glover et al.* 2213).

6.   **T. pellegriniana** *Keay* in B.J.B.B. 26: 189, fig. 59 (1956); F.W.T.A., ed. 2, 1: 708 (1958). Type: Nigeria, Obudu, Sonkwala Hills, Ijua, *Keay & Savory F.H.I.* 25029 (K, holo.!, P, iso.!)

Shrub or small tree up to 8 m. tall; older branchlets slender, closely longitudinally striate, often turning reddish brown or purple-black. Leaf-lamina elliptic or oblanceolate-elliptic, up to 12 × 5.5 cm., lower surface sparsely puberulous, apex acuminate, base cuneate; petiole up to 0.7 cm. long. Inflorescence a subsessile fascicle, usually on older branchlets, very rarely axillary, with up to 19 flowers, usually flowering when leafless; bracts 0.1 cm. long; pedicels up to 1.8 cm. long. Calyx 0.2 cm. long, sparsely sericeous-puberulous outside, lobed to the middle, lobes deltate. Petals ± 1.4 cm. long, sparsely and minutely puberulous outside. Staminal tube ± 1 cm. long, slightly expanded distally, glabrous inside for 0.25 cm. below insertion of the anthers; appendages fused to form a subentire frill completely concealing the anthers. Ovary 5–6-locular; style 1.2 cm. long; style-head thistle-shaped. Capsule orange, depressed-globose, ± 0.9 × 1 cm., shallowly sulcate, thinly woody, glabrous. Seeds ± 0.4 × 0.25 cm., black; aril orange or vermilion.

UGANDA. Bunyoro District: Budongo Forest, near Buchanana village, Mar. 1973, *Synnott* 1443!
DISTR. U 2; Nigeria, Cameroon, Central African Republic, Zaire, Sudan
HAB. At edges of lowland rain-forest; 1050 m.

SYN. *T. tisserantii* Pellegrin in Not. Syst. 9: 10 (1940), *nom illegit.* (gallice tantum descripta)

7. **T. holstii** *Gürke* in E.J. 19, Beibl. 47: 35 (1894) & in P.O.A. C: 231 (1895); Bak.f. in J.B. 41: 10 (1903); V.E. 3(1): 813 (1915); T.S.K., ed. 2: 103 (1936); T.T.C.L.: 321 (1949); I.T.U., ed. 2: 200 (1952); Staner & Gilbert in F.C.B. 7: 155 (1958); K.T.S.: 274 (1961); F. White & Styles in F.Z. 2: 311 (1963); Styles in E. Afr. Agric. Journ. 39: 421 (1974); Styles & F. White in Fl. Eth. 3: 481, fig. 124.2/9, 10 (1990). Type: Tanzania, W. Usambara Mts., near Lutindi, *Holst* 3392 (B, holo.†, G, K, M, P, iso.!)

Small tree up to 18 m. tall, more rarely a shrub, sometimes scrambling. Leaf-lamina lanceolate-elliptic, elliptic or oblanceolate-elliptic, usually less than 9 × 4 cm., lower surface almost glabrous except for a few hairs on the nerves and conspicuous tufts of short hairs in axils of the secondary nerves, apex rather suddenly, shortly and somewhat bluntly acuminate, base cuneate; petiole up to 0.9 cm. long. Inflorescence an axillary (1–)2–3(–4)-flowered cyme; peduncle 0.5–2.0 cm. long; bracts small, squamiform, ± 0.1 cm. long; pedicels (0.7–)2–4.3 cm. long, very slender. Calyx ± 0.25 cm. long, denticulate, puberulous, especially on the teeth and margin. Petals linear-spathulate, 1.4–2.5 × 0.2–0.3 cm., white, turning yellow with age, glabrous or sparsely and minutely puberulous near apex outside, or sometimes tomentellous. Staminal tube 0.8–1.8 cm. long, slightly expanded at the throat, glabrous immediately below the anthers, otherwise densely hairy in the upper two-thirds inside; appendages shorter than the anthers, entire and truncate, emarginate or lobed in upper half, usually fused for the greater part of their length to form a frill 0.1–0.15 cm. long beyond the insertion of the filaments. Ovary 5–6(–7)-locular, glabrous; style 1.5–3 cm. long; style-head usually ovoid, sometimes subglobose. Capsule depressed-globose, ± 0.7 × 1 cm., shallowly sulcate, thinly woody, glabrous. Seeds reddish black; aril red. Fig. 1/5, p. 9.

UGANDA. Acholi District: Imatong Mts., Aringa R. headwaters, Apr. 1945, *Greenway & Hummel* 7306!
KENYA. Trans-Nzoia District: Elgon, Feb. 1958, *Symes* 264!; Meru Forest, *Gardner* in F.D. 1326!; N. Kavirondo District: Kakamega Forest, Dec. 1956, *Verdcourt* 1670!
TANZANIA. Arusha District: Mt. Meru, Aug. 1954, *Hughes* 224!; Lushoto District: Mkuzi, Apr. 1953, *Drummond & Hemsley* 2111!; Lindi District: Rondo Plateau, Mchinjiri, Feb. 1952, *Semsei* 641!
DISTR. U 1; K 2–7; T 2, 3, 5–8; Zaire, Sudan, Ethiopia, Somalia, Malawi and Arabian Peninsula
HAB. In montane and mid-altitude forest, both in understorey and at edges, sometimes on stream banks or in abandoned cultivation; (700–)950–2500 m.

SYN. *T. abyssinica* A. Rich. var. *longipedicellata* Oliv. in F.T.A. 1: 331 (1868). Type: Ethiopia, Ankober, *Roth* 178 (K, holo.!)
T. usambarensis Gürke in P.O.A. C: 231 (1895); Bak.f. in J.B. 41: 10 (1903); V.E. 3(1): 815 (1915). Type: Tanzania, Tanga District, Amboni, *Holst* 2579 (B, holo.†, K, M, P, Z, iso.!)*
T. laxiflora C. DC. in Ann. Conserv. Jard. Bot. Genève 10: 127 (1907). Type: Tanzania, Morogoro District, Nguru Mts. [Ngourou], *Sacleux* 810 (P, holo.!)

NOTE. *Vesey-FitzGerald* 4440 (EA) from Ngurdoto Crater, northern Tanzania (T2) is anomalous in its more coriaceous, less distinctly acuminate leaves, shorter, stouter pedicels, tomentellous petals and unfused staminal appendages. It agrees however with *T. holstii* better than with any other species.

8. **T. stolzii** *Harms* in E.J. 46: 159 (1911); V.E. 3(1): 813 (1915); T.T.C.L.: 321 (1949). Type: Tanzania, Rungwe [Lungwe], *Stolz* 73 (B, holo.†, E, Z, iso.!)

---

* The specimen was not collected from the well-known locality on the coast. Engler, Gliederung der Vegetation von Usambara, in Abh. Preuss. Akad. Wiss. 1894: 51 (1894), mentions *T. usambarensis*, *nom. nud.*, as a component of upland dry evergreen forest (trockenere Tropenwald) in the W. Usambara Mts. The specimen is labelled June 1893. Holst made an excursion from Amboni to the W. Usambara Mts. from April to June 1893, ending at Dar es Salaam, which also appears on the printed label, but crossed out.

A shrub or small tree to 4.5 m. tall. Leaf-lamina elliptic to lanceolate-elliptic, up to 13.5 × 6.5 cm., coriaceous, glabrous except for scattered hairs on the midrib and lateral veins and tufts of hairs in their axils beneath, apex shortly acuminate, base usually cuneate; petiole up to 1 cm. long. Inflorescence axillary, a (2–)3(–4)-flowered cyme; peduncle up to 1.5 cm. long, glabrous, stout; pedicels 1.3–2 cm. long, glabrous, stoutish; bracts up to 0.3 cm. long, puberulous, deltate. Flowers white, fading to yellow. Calyx 0.4–0.6 cm. long, almost glabrous except for the densely puberulous, short and broadly obtuse teeth. Petals linear-spathulate, 2–2.8 × 0.4–0.5 cm., minutely puberulous towards the apex outside, otherwise glabrous. Staminal tube 1.6–2.2 cm. long, glabrous inside immediately below the insertion of the filaments, otherwise densely hairy with longish hairs in the remainder of the upper half inside; appendages fused for their entire length to form an irregularly lobed frill 0.35 cm. long beyond the insertion of the filaments, completely concealing the anthers. Ovary (4–)5(–6)-locular, glabrous; style 2.1–3 cm. long, glabrous; style-head ovoid or obovoid. Capsule depressed-globose, up to 1.2 × 1.4 cm., sulcate, thinly woody, glabrous. Seeds black; aril red.

TANZANIA. Mbeya District: Mbosi Circle, Boma River Estate, Jan. 1961, *Richards* 13929!; Rungwe District: Kyimbila, Jan. 1914, *Stolz* 2423!; Njombe District: Matamba, Feb. 1954, *Paulo* 259!
DISTR.   T 7; Malawi
HAB.   Montane rain-forest and montane riparian forest; 1400–1800 m.

NOTE.   *T. stolzii* is closely related to *T. holstii*, but constantly differs in its shorter, stouter pedicels and larger flowers, with the staminal tube more expanded at the throat; the staminal appendages are more or less completely fused and conceal the anthers.

9. **T. parvifolia** *Deflers* in Bull. Soc. Bot. Fr. 42: 301, fig. 6 (1895); Bak.f. in J.B. 41: 11 (1903); Schwartz, Fl. Trop. Arabia: 129 (1939); Blundell, Wild Fl. E. Afr.: 136, t. 5 (1987); Styles & F. White in Fl. Eth. 3: 481, fig. 124.2/11 (1990). Types: Yemen, El 'Areys Mountains (Bilad Fodhli), *Deflers* 990 (P, syn.!) & Wadi Moaden (Bilad Soubaihi), *Deflers* 1143 (P, syn.!, G, isosyn.!)

Rigidly branched shrub or small tree up to 4 m. tall; bark on older branchlets dark brown or blackish, finely striate. Leaves in fascicles on short lateral shoots of limited and slow growth, subsessile or with petiole up to 0.2 cm. long; lamina very variable, broadly obovate to linear-spathulate, 0.5 × 0.4–4.2 × 0.8 cm., glabrous, or with a few scattered hairs on the lower surface, apex obtuse or truncate, sometimes emarginate, base narrowly cuneate, decurrent almost to the base of the petiole. Flowers greenish cream fading yellow, solitary or 2–3(–4)-fasciculate among the leaves; pedicels up to 0.6 cm. long. Calyx 0.15–0.25 cm. long, denticulate, densely puberulose over whole surface. Petals greenish white, linear spathulate, 1.1–1.7 × 0.2–0.3 cm., sparsely puberulous outside. Staminal tube slightly curved, 0.8–1.2 cm. long, slightly expanded at the apex, with few to many longish hairs in the upper part inside; appendages alternating with the anthers, in pairs, the members of a pair either completely separate or united for up to half of length, up to 0.35 cm. long. Ovary 5–6-locular, glabrous; style 1.2–1.8 cm. long, glabrous; style-head ovoid. Capsule depressed-globose, ± 0.6 × 0.8 cm., shallowly sulcate, thinly woody, glabrous. Seeds black; aril red. Figs. 2/6 and 1/7, p. 9.

KENYA. Northern Frontier Province: Lomelo Mts., June 1970, *Mathew* 6739!; Meru District: 24 km. east of Isiolo, Shabele, Mar. 1963, *Bally* 12616!; Tana River District: Kora Rock, Apr. 1983, *Hemming* 83/42!
DISTR.   K 1, 3, 4, 7; Ethiopia, Somalia, Arabian Peninsula
HAB.   In *Acacia-Commiphora* bushland, especially around rock outcrops and on banks of seasonal watercourses; 260–1250 m.

SYN.   *T. lycioides* Bak. in K.B. 1895: 212 (1895), *nom. illegit., non T. lycioides* Baillon (1892). Types: Somalia, Golis Range, Dooloob, *Edith Cole* (K, syn.!) & *Lort Phillips* (K, syn.!)
    T. parviflora sensu Engl., V.E. 3(1): 813 (1915), sphalm.
    T. somaliensis Li & Chen in Acta Phytotaxonomica Sinica 22(6): 496 (1984), *nom. illegit.* Types: as above

NOTE.   Sometimes heavily browsed by camels, sheep and goats and producing a dwarf densely branched form.

10. **T. vogelii** *Benth.* in Hook.f., Niger Fl.: 253 (1849); F.T.A. 1: 330 (1868); C. DC. in A. & C. DC., Monogr. Phan. 1: 444 (1878); Bak.f. in J.B. 41: 11 (1903) & in J.L.S. 37: 132 (1905); Z.A.E.: 433 (1912); V.E. 3(1): 815, fig. 384 A–C (1915); C.F.A. 1: 319 (1951); I.T.U., ed. 2: 220 (1952); Staner & Gilbert in F.C.B. 7: 155 (1958); F.W.T.A., ed. 2, 1: 708 (1958); Styles in E. Afr. Agric. Journ. 39: 422 (1974). Type. Dioko, *Vogel* (K, holo.!)

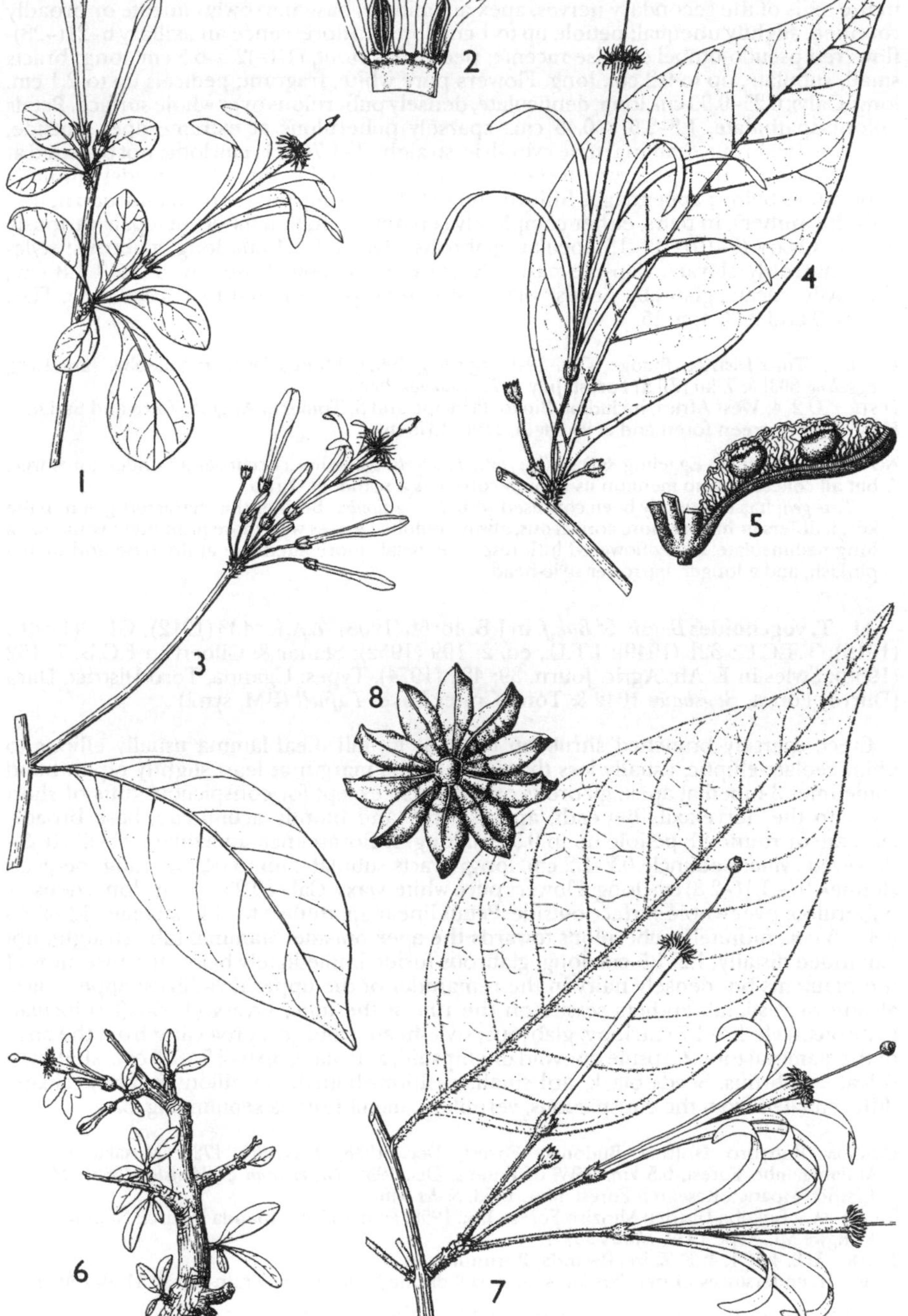

FIG. 2. *TURRAEA MOMBASSANA* subsp.*CUNEATA*—**1**, leaves and flowers, × ⁴⁄₅; **2**, details of staminal tube, × 4½. *T. VOGELII*—**3**, leaf and flowers, × ⁴⁄₅; **4**, details of fruit, × ⁴⁄₅. *T. WAKEFIELDII*—**5**, leaf and flowers, × ⁴⁄₅. *T. PARVIFOLIA*-**6**, habit and flowers, × ⁴⁄₅. *T. FLORIBUNDA*—**7**, leaf and flowers, × ⁴⁄₅; **8**, dehisced capsule, × ⁴⁄₅. 1, from *Conservator of Forests* 123; 2, from *Rogers* 409; 3, from *Louis* 14150; 4, from *Keay* s.n.; 5, from *Barbosa & de Lemos* 8548; 6, from *Azzaroti* s.n.; 7, from *Wild* 4333; 8, from *Watt & Brandwyck* 1095. Drawn by Julia Loken.

A climbing shrub up to 8 m. tall. Leaf-lamina ovate to elliptic or lanceolate-elliptic, up to 15 × 7 cm., usually glabrous beneath except for a few hairs on the nerves and tufts of hairs in the axils of the secondary nerves, apex acuminate, base narrowly cuneate or broadly rounded, slightly unequal; petiole up to 1 cm. long. Inflorescence an axillary 6–21(–28)-flowered pseudo-umbel or false raceme; peduncles stout, (1.1–)2.5–6.5 cm. long; bracts small, subulate, up to 0.2 cm. long. Flowers pure white, fragrant; pedicels up to 2.1 cm. long. Calyx 0.25–0.35 cm. long, denticulate, densely puberulous over whole surface. Petals oblong-spathulate, 1.8–2.5 × 0.45 cm., sparsely puberulous at extreme apex outside, otherwise glabrous. Staminal tube cylindric, straight, 1–1.7(–2.1) cm. long, not bearded at the throat, but densely hairy with short retrorse hairs in the upper half inside, minutely puberulous below; appendages filiform, up to 0.4 cm. long, longer than and alternating with the anthers, in pairs, either completely separate or the members of a pair variously united. Ovary (9–)10–12(–13)-locular, glabrous; style 1.8–2.4 cm. long, glabrous; style-head narrowly clavate. Capsule narrowly obovoid to subglobose, up to 3.5 × 1.8 cm., shallowly sulcate, glabrous. Seeds* black; aril and septum similar to *T. vogelioides*. Figs. 1/3, p. 9 and 2/3, 4, p. 15.

UGANDA. Toro District: Butogo, Oct. 1940, *Eggeling* 4062!; Mengo District: Entebbe, Jan. 1932, *Eggeling* 393! & Ziku [Zika] Forest, July 1975, *Katende* 2362!
DISTR. U 2, 4; West Africa, including Bioko, Principe and S. Tomé, to Angola, Zaire and Sudan
HAB. In evergreen forest and at its edges; 1190–1375 m.

NOTE. According to Eggeling & Dale (loc. cit.) *T. vogelii* sometimes occurs as an understorey tree, but all collectors who mention its habit record it as a scandent shrub.
    T. vogelii has frequently been confused with *T. vogelioides*. Besides the characters given in the key, it differs in having more coriaceous, always unlobed leaves with more prominent venation, a long-pedunculate, many-flowered inflorescence, petals more attenuate at the base and drying pinkish, and a longer, narrower style-head.

11. **T. vogelioides** *Bagsh. & Bak.f.* in J.B. 46: 56 (1908); Z.A.E.: 433 (1912); V.E. 3(1): 815 (1915); T.T.C.L.: 321 (1949); I.T.U., ed. 2: 199 (1952); Staner & Gilbert in F.C.B. 7: 152 (1958); Styles in E. Afr. Agric. Journ. 39: 422 (1974). Types: Uganda, Toro District, Dura [Durro] Forest, *Bagshawe* 1042 & Toro, Crater Lake, *Tufnell* (BM, syn.!)

Erect, sparsely branched shrub up to 3(–5) m. tall. Leaf-lamina usually elliptic to oblanceolate-elliptic, usually less than 12 × 6 cm., margin at least slightly repand and sometimes 3-lobed at apex, glabrous or almost so except for conspicuous tufts of short hairs in the nerve-axils beneath, apex shortly and bluntly acuminate, base broadly cuneate or rounded; petiole up to 0.5 cm. long. Inflorescence an axillary, (1–)2–4(–5)-flowered cyme; peduncle 0.2–1.2 cm. long; bracts subulate, up to 0.2 cm. long; pedicels slender, 0.5–1.1(–2.8) cm. long. Flowers pure white, waxy. Calyx 0.25–0.4 cm. long, densely puberulous over whole surface outside. Petals linear-spathulate to oblanceolate, 1.2–2.2 × 0.3–0.5 cm., minutely puberulous towards the apex outside. Staminal tube straight, not expanded distally, 1.2–1.5 cm. long, glabrous inside immediately below the insertion of the filaments but densely hairy in the remainder of the upper two-thirds; appendages absent or vestigial; anthers sessile on the rim of the tube. Ovary (4–)5–6(–7)-locular, glabrous; style 1.4–1.7 cm. long, glabrous; style-head ovoid, just projecting from the apex of the staminal tube. Capsule obovoid or ellipsoid, apiculate, up to 2.7 × 1.3 cm., shallowly sulcate, glabrous. Seeds black, aril small, cushion-shaped, vermilion (*in vivo*), clearly differentiated from the conspicuous, vermilion, membranous septum. Fig. 3.

UGANDA. Bunyoro District: Budongo Forest, Dec. 1938, *Loveridge* 172!; Masaka District: Malabigambo Forest, 6.5 km. SSW. of Katera, Oct. 1953, *Drummond & Hemsley* 4553!; Mengo District: Mpanga Research Forest, Mar. 1963, *Styles* 346!
TANZANIA. Bukoba District: Minziro Forest, Jan. 1958, *Procter* 776!; Mpanda District: Ntali R. below Kungwe Mt., Sept. 1959, *Harley* 9577!
DISTR. U 2, 4; T 1, 4; E. Zaire, Rwanda, Burundi
HAB. In understorey of riparian forest and well-drained and swampy rain-forest; 1150–1800 m.

SYN. [*T. heterophylla* sensu Eggeling & Dale, I.T.U., ed. 2: 200 (1952), *non* Sm.]

NOTE. *T. heterophylla* Sm. is only known from West Africa. It differs from *T. vogelioides* in its more coriaceous, more deeply lobed leaves and in having staminal appendages longer than the anthers.

* Fig. 2/4 was drawn from spirit material; in living material the seeds stand upright.

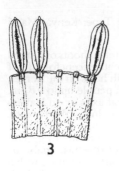

3

5

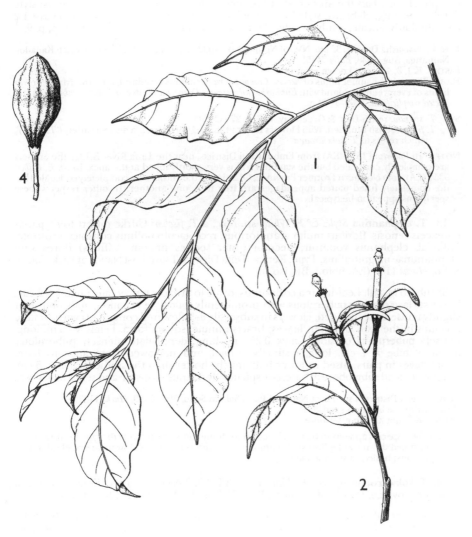

4

2

J. Loken

FIG. 3. *TURRAEA VOGELIOIDES*—**1**, leaves, × ½; **2**, flowers, × 1; **3**, details of staminal tube, × 5; **4**, unopened capsule, × 1; **5**, dehisced capsule, × 1⅕. All from cultivated plant (seed of *Harley* 9577) in Oxford Bot. Gdn. Drawn by Julia Loken.

12. **T. cornucopia** *Styles & F. White* in B.J.B.B. 59: 257 (1989) Type: Kenya, Masai District, 32 km. Kikuyu–Narok, *Verdcourt* 3552 (K, holo.!, EA, iso.!)

Shrub or small tree, up to 4 m. tall. Leaves in fascicles on short lateral shoots of limited growth; lamina ovate-elliptic, up to 7 × 3.5 cm., almost glabrous beneath except for tufts of hairs in the axils of the secondary nerves and a few scattered longish hairs on the nerves, apex rounded to bluntly subacuminate, base cuneate or rounded; petiole up to 0.8 cm. long. Flowers (rarely) solitary, or in a 2–10-flowered fascicle, in leaf-axils or on leafless spur shoots; pedicels up to 1.1 cm. long. Calyx 0.25–0.3 cm. long, denticulate, tomentellous. Petals linear-spathulate, 2.3–3.2 × 0.55 cm., glabrous or sparsely puberulous especially towards apex outside, greenish white. Staminal tube 1.8–2.8 cm. long, pure white, distinctly curved and gradually narrowed from the wide apex to the base, glabrous in the upper half inside, but with long retrorse hairs in the lower part, sometimes with a few scattered hairs outside; appendages fused for their entire length to form a frill which completely conceals the anthers (but see note below). Ovary 10-locular, glabrous; style 2.9–4.1 cm. long, glabrous; style-head ovoid. Capsule depressed-globose, up to 0.8 × 1.2 cm., shallowly sulcate, thinly woody, glabrous. Seeds black with red aril. Fig. 1/6, p. 9.

KENYA. Naivasha District: 32 km. NW. of Ngong, Nov. 1932, *C. G. Rogers* 32!; Masai District: Kajiado–Namanga, Aug. 1938, *Bally* 7445!
DISTR. **K** 1, 3, 4, 6; not known elsewhere
HAB. Towards the upper limits of *Acacia–Commiphora* deciduous bushland, and towards the lower limits of evergreen bushland with *Euphorbia kibwezensis, Dracaena ellenbeckii, Vepris samburensis* and *Teclea simplicifolia*; 1200–1830 m.

SYN. *T. sp. nov.* sensu Dale & Greenway, K.T.S.: 275 (1961)
     T. fischeri sensu Blundell, Wild Flowers E. Afr.: 135, t. 63 (1987) excl. 'a distinct variety ('*T. sp. B*')', which refers to *T. kokwaroana*

NOTE. In *Robertson* 1614 (EA) from Embu/Kitui Districts, near the Tana River Bridge the anthers are completely concealed but the staminal frill is very shortly lobed at the apex. In *M. E. Gilbert* 5356 (EA) from Northern Frontier Province, Ol Lolokwe the anthers are completely concealed but the completely fused paired appendages are free from one another. In other respects these specimens appear to be typical.

13. **T. elephantina** *Styles & F. White*, sp. nov., a *T. fischeri* Gürke differt foliis parvis coreaceis venosis, floribus multo grandioribus, receptaculo pollinis conspicue rostrato. Folia ab elephantis vorantur; necnon (inter species proxime affines) flores sunt elephantinae magnitudinis. Type: Kenya, Teita District, Tsavo East National Park, Sala to Sobo, *Hucks* 1187 (EA, holo.!, BR, iso.!)

Shrub 2.6 m. tall. Leaf-lamina coriaceous, ovate to broadly lanceolate, up to 5 × 2 cm., lower surface densely puberulous with prominently reticulate venation, apex obtuse or slightly retuse, base rounded. Flowers in subsessile 3–5-flowered congested cymes or false racemes in the axils of fallen leaves; bracts minute, squamiform. Petals ± 4 cm. long, densely puberulous outside. Calyx 0.25 cm. long, denticulate, densely puberulous. Staminal tube ± 4 cm. long, distinctly curved, and narrowed from apex to base; appendages in pairs, fused in lower half, much shorter than the anthers. Style ± 5 cm. long; style-head subglobose at base, conspicuously rostrate. Capsule and seeds unknown.

KENYA. Teita District: Tsavo East National Park, Sala to Sobo, Nov. 1969, *Hucks* 1187!
HAB. In sparse bush on sandstone
DISTR. **K** 7; not known elsewhere

NOTE. This species appears to be rare. Its collector found only one plant at the type locality. It had been pruned down by elephants; also among its closest relatives, its flowers are of elephantine proportions; hence its specific name.

14. **T. kokwaroana** *Styles & F. White* in B.J.B.B. 59: 258 (1989). Type: Kenya, Northern Frontier Province, Dandu Mt., *Gillett* 12577 (K, holo.!, BM, EA, FT, PRE, W, iso.!)

Shrub or small tree up to 6 m. tall. Leaf-lamina broadly ovate, broadest just above the base, up to 6 × 3.8 cm., lower surface densely setulose at first, more sparsely so with age, apex rounded to subacute, base subtruncate to shallowly cordate; petiole 0.8–1.2 cm. long. Inflorescence a 3–12-flowered subsessile fascicle terminating short shoots of limited growth or in axils of fallen leaves, rarely axillary. Flowers creamy white, precocious or with the leaves; pedicels ± 1 cm. long. Calyx ± 0.35 cm. long, lobed almost to the middle, tomentellous. Petals linear-spathulate, 2.8–3.2 × 0.4–0.5 cm., sparsely puberulous outside.

Staminal tube 2.2–2.4 cm. long, distinctly curved and gradually narrowed from apex to base, glabrous at throat inside but densely hairy in the lower half; appendages in pairs, alternating with and ± equalling the anthers, free to the base or almost so. Ovary 10-locular, densely puberulous; style ± 3 cm. long, puberulous in lower half; style-head ovoid. Capsule depressed-globose, up to 0.8 × 1.2 cm., shallowly sulcate, thinly woody, densely puberulous. Seeds black; aril red.

KENYA. Northern Frontier Province: Dandu Mt., July 1952, *Gillett* 13615! & Marsabit–Moyale road near Gof Choba Hill, Mar. 1963, *Bally* 12570!
DISTR. **K** 1; not known elsewhere
HAB. *Commiphora, Acacia* scrub with scattered larger trees of *Delonix, Terminalia* and *Gyrocarpus;* sometimes in rocky places; 775 m.

SYN. *T. sp. 2* sensu Dale & Greenway, K.T.S.: 276 (1961), pro parte excl. *Gillett* 12526

15. **T. fischeri** *Gürke* in E.J. 14: 308 (1891) & in P.O.A. C: 231 (1895); Bak.f. in J.B. 41: 11 (1903); V.E. 3(1): 815 (1915); Milne-Redh. in K.B. 1936: 475 (1936); T.T.C.L.: 320 (1949); I.T.U., ed. 2: 200 (1952); Styles in E. Afr. Agric. Journ. 39: 421 (1974). Types: Tanzania, without precise locality, *Fischer* 93 (B, syn.†, FHO, fragment, isosyn.!) & *Fischer* 94 (B, syn.†)

Shrub or much-branched deciduous tree to 9 m. tall. Leaf-lamina ovate, ovate-elliptic, up to 9 × 5.5 cm., lower surface softly pubescent or hairs confined to nerves and nerve-axils or completely glabrous, apex acute to shortly acuminate, petiole up to 1.2 cm. long. Inflorescence a (3–)4–12-flowered fascicle terminating leafless lateral shoots of limited growth; bracts ± 0.2 cm. long, squamiform. Flowers creamy white, fragrant, usually precocious. Pedicels 0.6–1.5 cm. long. Calyx 0.2–0.35 cm. long, denticulate, densely puberulous over whole surface. Petals linear-spathulate, 2–3.2 × 0.3–0.5 cm., sparsely to densely puberulous outside. Staminal tube 1.6–2.4 cm. long, distinctly curved and gradually narrowed from apex to base, glabrous at throat inside but hairy in lower half; appendages alternating with the anthers in pairs, shorter than the anthers, the members of a pair fused from one-quarter to three-quarters of length. Ovary (7–8–)9–10-locular, densely puberulous; style 2.5–3 cm. long, puberulous in lower half; style-head ovoid, its base exserted 0.8–1.4 cm. beyond the staminal-tube. Capsule depressed-globose, ± 0.8 × 1.3 cm., shallowly sulcate, thinly woody, sparsely to densely puberulous. Seeds black; aril red. Fig. 1/4, p. 9.

UGANDA. Acholi District: Mt. Rom, Dec. 1935, *Eggeling* 2396!; Karamoja District: Toror Mt., Dec. 1958, *J. Wilson* 540!
TANZANIA. Shinyanga District: Usule, Nov. 1938, *Koritschoner* 1643!; Musoma District: Seronera, Oct. 1960, *Greenway* 9731!; Mpwapwa District: Mfutu R., Aug. 1933, *B.D. Burtt* 4761!
DISTR. **U** 1; **T** 1, 2–5; Zimbabwe (subsp. *eylesii* (Bak.f.) Styles & F. White)
HAB. Bushland and scrub forest, mainly on rock outcrops; 1200–1850 m.

NOTE. The roots are used in cases of tapeworm and other intestinal worms but not hookworm (*Koritschoner* 1643).

16. **T. mombassana** *C. DC.* in A. & C. DC., Monogr. Phan. 1: 439, fig. 6/4 (1878); Gürke in P.O.A. C: 231 (1895); Bak.f. in J.B. 41: 10 (1903); C. DC. in Ann. Conserv. Jard. Bot. Genève 10: 125 (1907); V.E. 3(1): 813, fig. 384D,E (1915); T.S.K., ed. 2: 103 (1936); T.T.C.L.: 320 (1949); K.T.S.: 274 (1961); F. White & Styles in F.Z. 2: 312 (1963); Blundell, Wild Fl. E. Afr.: 136, t. 64 (1987); Styles & F. White in Fl. Eth. 3: 481, fig. 124.2/1–3 (1990). Type: Kenya, near Mombasa, *Hildebrandt* 1973 (G, holo.!, BM, K, M, P, iso.!)

Shrub or subshrub up to 5 m. tall, sometimes scrambling. Leaves mostly in fascicles; lamina very variable in shape and size, rhombic or rhombic-spathulate, but outline occasionally irregular or shallowly 3-lobed, up to 6.3 × 2.6 cm. but usually much smaller, lower surface usually glabrous, except for a few scattered hairs on the nerves and conspicuous tufts of short hairs in the nerve-axils, apex subacuminate to rounded or emarginate, base narrowly cuneate, decurrent almost to the insertion of the petiole; petiole very short or absent. Flowers solitary or 2–6-fasciculate among the leaves on spur-shoots or (more rarely) in leaf-axils on long shoots; bracts very small, 0.1 cm. long, subulate; pedicels 0.3–1 cm. long. Calyx 0.25–0.45(–0.6) cm. long, lobed to ± the middle; lobes narrowly deltate (very rarely subulate), puberulous especially on the teeth. Petals linear-spathulate, 3.2–5.8(–6) × 0.4–0.5 cm., glabrous on both surfaces. Staminal tube 2–4.3(–5.4) cm. long, cylindric, glabrous inside in the upper half, but with longish retrorse hairs in the lower half; appendages entire or laciniate, in pairs, 0.3–0.4 cm. long, usually free, occasionally the members of a pair fused near the base, glabrous outside. Ovary 5(–6)-locular, densely setulose-puberulous; style 4–5.5 cm. long; style-head conical-cylindric from a broad base. Capsule depressed-globose, ± 0.5 × 1 cm., very shallowly sulcate, thinly woody, glabrous. Seeds red; aril small, white, contrasting strikingly with the seed, but completely hidden unless the seed is removed.

a. subsp. **mombassana**

Leaves 3–6.2 cm. long, apex distinctly subacuminate. Staminal tube 2–2.8 cm. long.

KENYA. Kwale District: Shimba Hills, Mwele Mdogo Forest, Feb. 1953, *Drummond & Hemsley* 1108!; Kilifi District: Sokoke, *R.M. Graham* 1947!; Lamu District: Utwani Forest, Dec. 1956, *Rawlins* 230! TANZANIA. Tanga District: Kange Estate, Jan. 1952, *Faulkner* 881!; Uzaramo District: Pugu Forest Reserve, June 1954, *Semsei* 1722!; Iringa District: Image Mt., Mar. 1954, *Carmichael* 382!; Zanzibar I., Kiungani, Feb. 1929, *Greenway* 1313! DISTR. K 7; T 1, 3, 4, 6–8; Z; Malawi; replacing subsp. *cuneata* to the east and south, in moister forest and mainly at lower altitudes. HAB. Evergreen forest, both inside and at edge; also in coastal scrub and in secondary forest colonizing abandoned cultivation; 0–1675 m.

SYN. ?*Pittosporum spathulifolium* Engl. in E.J. 43: 372 (1909). Type: Tanzania, Lindi, *Busse* 2995 (B, holo.†)

b. subsp. **cuneata** (*Gürke*) *Styles & F. White* in B.J.B.B. 59: 258 (1989). Tanzania, without precise locality, *Fischer* 266 (B, holo.†); Kenya, Kitui District, Galunka, *Kassner* 830 (BM, neo., K, isoneo.!) —see Styles & F. White (1989)

Leaves (0.6–)1.5–3.3 cm. long, apex usually rounded to subacute, rarely emarginate. Staminal tube 3–4.3 cm. long. Fig. 2/1, 2, p. 15.

KENYA. Northern Frontier Province: Moyale, July 1952, *Gillett* 12937!; Naivasha District: Kedong, Mt. Margaret Estate, June 1940, *Bally* 1082!; Nairobi, Thika Road House, Dec. 1950, *Verdcourt* 398! TANZANIA. Masai District: Longido Mt., Jan. 1936, *Greenway* 4375!; Pare District: Suji, Aug. 1928, *Haarer* 1560!; Kondoa District: Kinyassi scarp, Feb. 1928, *B.D. Burtt* 1340! DISTR. K 1, 3, 4, 6, 7; T 1–3, 5; Ethiopia; replacing subsp. *mombassana* in drier country to the north and west, mainly at higher altitudes. HAB. Mainly in dry forest, bushland and thicket, especially on rocky slopes, sometimes in grassland; 1525–2225 m.

SYN. *T. cuneata* Gürke in P.O.A. C: 231 (1895).
     Pittosporum jaegeri Engl. in E.J. 43: 372 (1909). Type: Tanzania, Mbulu District, near Mt. Hanang [Gurui], *Jaeger* 286 (B, holo.†)
     Turraea mombassana C. DC. var. *cuneata* (Gürke) Engl., V.E. 3(1): 813 (1915)

c. subsp. **schliebenii** (*Harms*) *Styles & F. White*, comb. et stat nov. Type: Tanzania, Uluguru Mts., *Schlieben* 3416 (B, holo.†, P, lecto.!, here designated; BR, isolecto.!)

Shrub 2–3 m. tall. Closely resembling subsp. *mombassana* in its relatively large subacuminate leaves but differing in its much larger flowers with the calyx ± 0.6 cm. long, with subulate lobes, and the staminal tube ± 6 cm. long.

TANZANIA. Morogoro District: Uluguru Mts., without precise locality, Feb. 1933, *Schlieben* 3416! DISTR. T 6; not known elsewhere. HAB. Montane forest; 1500 m.

SYN. *T. schliebenii* Harms in N.B.G.B. 11: 1070 (1934); T.T.C.L.: 320 (1949)

DISTR. (of species as a whole). K 1, 3, 4, 6, 7; T 1–8; Z; Ethiopia, Malawi; a related species (*T. obtusifolia* Hochst.) occurs in southern Africa.

NOTE. Subsp. *mombassana* and subsp. *cuneata* show almost complete ecogeographical replacement, and most specimens are easy to identify. Their distinctive features however break down sufficiently often to make specific rank inappropriate. The following specimens (among others) are variously intermediate or otherwise anomalous: *Greenway & Kanuri* 15356 and *Procter* 3788 (Kilosa District), *Procter* 3135 (Katonga Gorge), and *Harris* 2777 (Dar es Salaam) from Tanzania; and *Rawlins* 230 (lacking staminal appendages, Utwani Forest) and *Williams* 542 (Nairobi, Arboretum, presumably cultivated).

17. **T. wakefieldii** *Oliv.* in Hook., Ic. Pl. 15, t. 1489 (1885); Gürke in P.O.A. C: 231 (1895); Bak.f. in J.B. 41: 12 (1903); V.E. 3(1): 815 (1915); T.T.C.L.: 320 (1949); K.T.S.: 275 (1961); F. White & Styles in F.Z. 2: 314, fig. 61B (1963). Type: Kenya, near Mombasa, *Wakefield* (K, holo.!)

Shrub or small tree up to 7 m. tall, sometimes scrambling. Leaves sometimes crowded at the ends of slow-growing short shoots but usually widely spaced on long shoots; lamina coriaceous, ovate to lanceolate-elliptic, usually less than 7.5 × 4.5 cm., lower surface glabrous except for tufts of hairs in the axils of the secondary nerves, apex shortly and bluntly subacuminate, base cuneate; petiole up to 1.5 cm. long. Inflorescence subfasciculate, 1–6-flowered, sessile or subsessile; bracts squamiform, 0.2 cm. long; pedicels 0.7–1.8 cm. long. Flowers white fading to creamy yellow. Calyx 0.4 cm. long, denticulate or shallowly lobed, densely puberulous. Petals linear, (4.5–)5–6.6 × 0.15–0.2 cm., puberulous outside. Staminal tube (3.8)–4.5–5 cm. long, cylindric, glabrous in the upper part inside, but with a few longish irregular retrorse hairs in the lower part; appendages in pairs alternating with the anthers, free or slightly fused at the base, 0.35–0.4 cm. long. Ovary 10–12-locular, tomentellous; style 5–6.5 cm. long, glabrous; style-head conical-cylindric, tapering from a broad base. Capsule up to 2.5 × 1.5 cm., similar to *T. floribunda* but less deeply ribbed and sulcate, fulvous-tomentellous and with thinner valves. Seeds red; aril white. Fig. 2/5, p. 15.

KENYA. Teita District: near Voi, Apr. 1955, *J. Wilson* 53!; Kwale District: near Taru, Sept. 1953, *Drummond & Hemsley* 4187
TANZANIA. Uzaramo District: near Dar es Salaam, Dec. 1936, *Vaughan* 2467!; Lindi District: Mlinguru, Dec. 1934, *Schlieben* 5736!
DISTR. K 7; T 3, 6, 8; Somalia, Mozambique
HAB. Coastal forest and secondary forest; near sea-level

SYN. *T. kirkii* Bak.f. in J.B. 41: 13 (1903). Type: Somalia, Kismayu, *Kirk* (K, holo.!)

NOTE. A juvenile specimen (*Robertson* s.n., FHO) grown from seed in a garden in Malindi, Kenya, has deeply pinnately lobed leaves. *Rodgers & Hall* 2356 (K) collected from a 1m. tall individual on Mafia I. appears to be conspecific.

18. **T. floribunda** *Hochst.* in Flora 27: 297 (1844); C. DC. in A. & C. DC., Monogr. Phan. 1: 445, fig. 6/4 (1878); Bak.f. in J.B. 41: 12 (1903); Z.A.E. 433: (1912); V.E. 3(1): 815 (1915); Stapf in Bot. Mag. 149, t. 8984 (1923); T.T.C.L.: 320 (1949); I.T.U., ed. 2: 200 (1952); Staner & Gilbert in F.C.B. 7: 154 (1958); K.T.S.: 274 (1961); F. White & Styles in F.Z. 2: 314, fig. 61E (1963); Killick in Fl. Pl. Afr., t. 1499 (1967); Styles in E. Afr. Agric. Journ. 39: 421 (1974); F. White in Bothalia 16: 151 (1986); F. White & Styles in F.S.A. 18: 45, fig. 11/4 (1986). Type: South Africa, Natal, Umlaas R., *Krauss* 342 (? B, holo.†, G, K, M, W, iso.!)

Deciduous shrub or small tree up to 10(–15) m. tall, sometimes scrambling. Leaf-lamina ovate to lanceolate, usually less than 15 × 8 cm., densely setose when young, later more sparsely so except on the nerves, apex acuminate, base subtruncate, rounded or broadly cuneate; petiole up to 2 cm. long. Inflorescence a (2–)3–18-flowered false raceme, or flowers subfasciculate. Flowers very variable in size, appearing with or just before the leaves, fragrant; peduncle up to 0.7 cm. long; bracts 0.2–0.4 cm. long. Calyx 0.3–0.6 cm. long, denticulate or shallowly lobed, densely puberulous or tomentellous. Petals linear-spathulate, 3.2–9.4 × 0.3–0.5 cm., greenish white, sparsely puberulous outside. Staminal tube 2.4–10.8 cm. long, pure white, cylindric, glabrous inside; appendages in pairs, alternating with the anthers, up to 0.8 cm. long, glabrous outside. Ovary usually 10- or 12-locular, glabrous or sparsely puberulous; style 3.2–12 cm. long, glabrous; style-head globose, its base exserted 0.6–1.8 cm. beyond the staminal tube. Capsule usually obovoid-cylindric, more rarely globose or depressed-globose, up to 2.5 × 1.5 cm., deeply ribbed and sulcate, glabrous, dark brown or black, the valves very thick (± 0.5 cm.), woody, divergent like the rays of a starfish. Seeds orange, contrasting with the whitish aril which covers one third to one half of the seed and is visible on its flanks. Fig. 2/7, 8, p. 15.

UGANDA. Bunyoro District: Budongo Forest, Mar. 1933, *Eggeling* 1276!; Mbale District: Budama, Sept. 1939, *Dale* U40!
KENYA. Kericho District: Aitibu, Sotik, July 1957, *Dale* 1002!; Kwale District: Mrima Hill, Nov. 1978, *Brenan et al.* 14622! & Shimba Hills, *Battiscombe* 111!
TANZANIA. Pangani District: Bushiri Estate, Mar. 1950, *Faulkner* 526 & Hale – Makinyumbe, Pangani R., July 1953, *Drummond & Hemsley* 3136!; Morogoro, Merera stream, May 1946, *Wigg* in FH 1381!; Rungwe District: Kyimbila, *Stolz* 1641!
DISTR. U 1–4; K 3, 5–7; T 2, 3, 6–8; Sudan, Zaire, Rwanda, Burundi, Malawi, Mozambique, Zimbabwe, South Africa (Natal and Cape Province); widely grown for ornament in the tropics and in conservatories in temperate countries
HAB. In evergreen forest, including riverine forest, especially at the edges, and in secondary forest and on abandoned cultivation; 100–2150 m.

SYN. *T. floribunda* Hochst. var. *macrantha* Oliv. in H.H. Johnston, The Kilimanjaro Expedition, Appendix to Chapter 17: 339 (1886), *nom. nud.*

T. kaessneri Bak.f. in J.B. 41: 13 (1903); T.S.K., ed. 2: 103 (1936); T.T.C.L.: 320 (1949); K.T.S.: 274 (1961). Type: Kenya, Kwale District, Shimba Hills, 460 m., *Kassner* 384 (BM, holo.!)

NOTE. The bark is used as an emetic (*Koritschoner* 1494) and the roots and bark as a purgative (*Koritschoner* 548).

The leaves of *T. floribunda* and *T. abyssinica* are similar in shape, and the two species have sometimes been confused in herbaria. Those of the former are usually less attenuate at the apex and lack axillary tufts of hairs on the lower surface. The infructescence of *T. abyssinica* has a longer peduncle with foliaceous bracts and fruits with only 5 (thinly woody) valves.

In South Africa (Bothalia 16: 151 (1986)) the flowers of *T. floribunda* are visited and presumably pollinated by hawkmoths. The length of flower is very variable (staminal tube 2.4–10.8 cm. long). Long-flowered variants (*T. kaessneri* Bak. f.) are concentrated in SE. Kenya and NE. Tanzania. Variation in this feature, however, is continuous and is not apparently correlated with others.

2. MELIA

L., Sp. Pl.: 384 (1753) & Gen. Pl., ed. 5: 182 (1754); Pennington & Styles in Blumea 22: 463 (1975)

Trees. Indumentum of simple, glandular and stellate hairs. Leaves 2–3-pinnate; leaflets subentire to deeply serrate or crenate. Flowers ♀ and ♂ on the same individual (plant andromonoecious). Calyx lobed almost to the base; lobes 5(–6), lanceolate, imbricate. Petals 5(–6), free, imbricate. Staminal tube narrowly cylindric, slightly expanded at the mouth; anthers 10, shortly apiculate; appendages alternating with or opposite to the anthers. Disk annular, free, surrounding the base of the ovary. Ovary 4–8-locular; locules with 2 superposed ovules; style-head scarcely wider than the style, with 4–8 erect stigmatic lobes. Fruit 3–8-locular drupe; locules usually 1-seeded.

3 species, 2 of which are indigenous to Africa; the third, *M. azedarach* as a wild plant extends from India to Australia — it is widely planted, and locally naturalized, in the tropics and subtropics.

Bark smooth; leaflets serrate; inflorescence lax; staminal tube
    violet; drupe up to 1.5(–2) cm. long . . . .   1. *M. azedarach*
Bark rough and deeply fissured; leaflets entire or subentire;
    inflorescence congested; staminal tube white; drupe
    usually 3–4 cm. long . . . .   2. *M. volkensii*

1. **M. azedarach** *L.*, Sp. Pl.: 384 (1753); F.T.A. 1: 332 (1868); C. DC. in A. & C. DC., Monogr. Phan. 1: 451 (1878); T.T.C.L.: 317 (1949); C.F.A. 1: 317 (1951); Staner & Gilbert in F.C.B. 7: 173 (1958); F. White & Styles in F.Z. 2: 315 (1963); Styles in E. Afr. Agric. Journ. 39: 416 (1974); Mabberley in Gard. Bull., Singapore 37: 55 (1984); F. White in Bothalia 16: 154 (1986); F. White & Styles in F.S.A. 18(3): 49, fig. 14 (1986); Styles & F. White in Fl. Eth. 3: 485 (1990). Lectotype, chosen by Mabberley (1984): Holland, de Hartecamp, cult. *Hort. Cliff.* 161.1 (BM, lecto.)

Medium-sized rapidly growing short-lived tree up to 15 m. tall, sometimes flowering as a shrub; bark grey-brown, smooth. Leaves 2(–3)-pinnate; petiole and rhachis up to 40 cm. long; leaflets opposite or subopposite, ± lanceolate, up to 5.5 × 2.5 cm., apex acuminate or subacuminate, base asymmetric, margin rather deeply crenate or serrate, lower surface sparsely puberulous, glabrescent, petiolules up to 0.7 cm. long. Inflorescence a large many-flowered axillary cymose panicle; flowers sweetly scented. Calyx 0.25 cm. long, densely stellate puberulous. Petals spathulate, up to 0.8 × 0.3 cm., pale lilac, sparsely puberulous outside with simple hairs, glabrous inside except for a few hairs towards apex. Staminal tube up to 0.7 cm. long, dark purple, glabrous outside, hairy inside, especially in the upper half; appendages ± 0.1 cm. long. Ovary less than 0.1 cm. in diameter, 5(–7)-locular; style 0.45 cm. long. Drupe up to 2 × 1.5 cm., pale yellow; putamen with ± 5 longitudinal ridges and indistinct apical and basal depressions. Fig. 4.

UGANDA. Acholi District: 48 km. Gulu–Pakwach, Dec. 1962, *Styles* 260!; Bunyoro District: Budongo Forest Reserve, Nyabeya [Nyabyeya] Arboretum, Sept. 1962, *Styles* 40!; Mengo District: Entebbe Botanic Gardens, Sept. 1962, *Styles* 16!

KENYA. Baringo District: Loruk, Nov. 1983, *Kimunya* 43!; Nairobi, National Museum Grounds, *Mwangangi* 1800!; Kilifi District: Jilore Forest Station, *Perdue & Kibuwa* 10118!

FIG. 4. *MELIA AZEDARACH*—1, leaf and inflorescence, × ⅔; **1a**, half flower, vertical section, × 3⅓; **1b**, details of staminal tube, × 8; **1c**, fruits, × ⅔; **1d–1h**, stages in germination, × ⅔; 1–1b, from *Taylor* 241; 1c, after Watt & Breyer-Brandwijk, The medicinal and poisonous plants of southern and eastern Africa, ed. 2 (1962); 1d–1h, after Troup, The silviculture of Indian trees 1 (1921). Drawn by Rosemary Wise.

FIG. 5. *MELIA VOLKENSII*—**1**, flowering shoot, × ½; **2**, shoot with flowers and fruits, × ½; **3**, half flower, vertical section, × 5⅖; **4–6**, three views of putamen, × ⅗₀. 1, from *Willan* 428; 2, from *Taylor* 265; 3, from *Verdcourt* 2350; 4–0, from *Gillett* 21821. Drawn by Julia Loken.

TANZANIA. Kigoma District: Gombe Stream Reserve, Feb. 1964, *Pirozynski* 424!; Mpwapwa District: Kongwa Ranch, Feb. 1966, *Leippert* 6271!; Uzaramo District: Dar es Salaam, July 1969, *Mwasumbi* 10561!

DISTR. U 1–4; K 3, 4, 7; T 4–6; native of the eastern tropics from India to Australia; widely planted in the tropics and subtropics

HAB. Planted for ornament, building, shade, firewood and medicine, and locally naturalized, occurring in secondary grassland and thicket and on waste ground; 0–2150 m.

SYN. *M. dubia* Cav., Dissert.: 364 (1789); Dale, Introd. Trees Uganda: 50 (1953). Type: of uncertain provenance, *Sonnerat* (P-LAM, holo.)

NOTE. The bluish-flowered cultivars of *Melia azedarach* originated in Asia (Persian Lilac, China Berry). They are now one of the more familiar sights in the landscape of East Africa, where they are widely cultivated because of their beauty and many uses.

Locally, *M. azedarach* is naturalized but in East Africa, unlike southern Africa, it does not seem to have become a serious pest.

The fruit of *Melia* has long been known to be poisonous to some animals including the pig, and the deaths of children have been reported. The endocarps, which can easily be pierced at the ends, are used for beads. Poles of Persian Lilac are used in house building because of their durability. The more rapidly growing and taller 'White Cedar' variant (*M. dubia*) has been planted for forestry purposes in Uganda.

2. **M. volkensii** *Gürke* in P.O.A. C: 231 (1895); T.T.C.L.: 317 (1949); K.T.S.: 269 (1961); Styles & F. White in Fl. Eth. 3: 485; fig. 124.4 (1990). Types: Kenya, Machakos/Kitui District, Ukamba, *Hildebrandt* 2630 & Tanzania, Moshi District, Lake Chala [Dschallasee], *Volkens* 1786 & Lushoto District, Mazinde [Masinde], *Fischer* 161 (B, syn. †). Neotype, chosen here: Kenya, Meru District, Isiolo, *Gillett* 15168 (K, neo.!, EA, isoneo.!)

Tree up to 15(–20) m. tall; crown often with a browse-line produced by giraffe; bark with pronounced vertical fissures. Leaves up to 35 cm. long, less divided than in *M. azedarach;* leaflets nearly always entire (rarely a few leaflets have a single basal lobe or a few shallow teeth). Inflorescence congested, up to 12 cm. long, axillary and on older branchlets. Drupe 3–4 cm. long; endocarp very thick and bony, terete, with a star-like, 5-lobed apical depression and a rose-like, 5-lobed basal depression; fertile locules 2. Fig. 5.

KENYA. Northern Frontier Province: Mt. Lolokwe (Ol Donyo Sabachi), Mar. 1978, *M. Gilbert* 5025!; Machakos District; near Kanga Station, Feb. 1966, *Gillett & Burtt* 17048!; Masai District: Tsavo West National Park, near Kilaguni Lodge, July 1966, *V.C. Gilbert* C13!

TANZANIA. Pare District; Ibaya, Mar. 1965, *Harris* 77! & Same–Kisiwani, Aug. 1951, *Greenway* 8584!; Lushoto District; Mkomazi–Hedaru, Aug. 1967, *Procter* 3717!

DISTR. K 1, 4, 6, 7; T 3; Ethiopia, Somalia

HAB. An emergent in *Acacia, Commiphora* deciduous bushland; sometimes fringing seasonal watercourses and on rock outcrops; 350–1675 m.

NOTE. The fruits of *M. volkensii* are said to be eaten by giraffe and the stones appear in their droppings (Bally, pers. comm. reported by *Gillett* 21824), and also by goats. Foresters formerly experienced difficulty in growing this tree from seed. Cutting away part of the endocarp assists germination and propagation by root suckers is recommended. It is one of the most prized trees in the dry areas east of Mt. Kenya (W. Teel, A pocket directory of trees and seeds in Kenya (1984)). The timber which works easily and planes well has a marked ribbon figure when radially sawn (not dissimilar from that of *Khaya anthotheca*).

Forms of *Melia azedarach* with subentire leaflets can be confused with *M. volkensii*, but they are much more suddenly attenuated at the apex as shown in fig. 4.

3. AZADIRACHTA

A. Juss. in Mém. Mus. Nat. Hist. Nat., Paris 19: 220, t. 2/5, under pl. 13 (1830)*; Pennington & Styles in Blumea 22: 464 (1975)

Trees. Indumentum of simple hairs. Leaves pinnate; leaflets coarsely serrate. Flowers ♂ and ⚥ on the same individual (plant andromonoecious). Calyx 5-lobed to the lower half; lobes imbricate. Petals 5, free, imbricate. Staminal tube cylindrical, slightly expanded at the mouth, terminated by 10, rounded, truncate, emarginate or bilobed appendages,

---

* The precise date of publication is uncertain (see Stafleu & Cowan, Taxonomic literature, ed. 2, 2: 476 (1979)).

Rosemary Wise.

FIG. 6. *AZADIRACHTA INDICA*—1, inflorescences, × ⁹/₁₀; 2, half flower, vertical section, × 6¾; 3, apex of staminal tube × 6¾; 4, fruiting shoot, × ⁹/₁₀; 5–7, stages in germination, × ²/₃. 1–4, from *Sharma* s.n.; 5–7, after Troup, The silviculture of Indian trees 1 (1921). Drawn by Rosemary Wise.

which are sometimes partly united to form a frill. Disk annular, fused to the base of the ovary. Ovary 3-locular; locules with 2 collateral ovules; style-head scarcely expanded, ending in 3 acute, partly fused stigmatic lobes. Fruit a 1(–2)-seeded drupe with a thin cartilaginous endocarp.

Two species indigenous to the Indo-Malesian region. *A. indica*, the Neem Tree, is widely planted and locally naturalized in the drier tropics.

A. indica *A. Juss.* in Mém. Mus. Nat. Hist. Nat., Paris 19: 221 (1830); T.T.C.L.: 314 (1949); Dale, Introd. Trees Uganda: 11 (1953); F.W.T.A., ed. 2, 1: 708 (1958); Styles in E. Afr. Agric. Journ. 39: 408 (1974); F. White in Bothalia 16: 155 (1986). Styles & F. White in Fl. Eth. 3: 485, fig. 124.5 (1990). Type: from India

Evergreen tree, usually less than 15 m. tall; crown rounded, widespreading ; bole straight, short, stout; bark brown, rough and fissured. Leaves up to 40 cm. long; leaflets ovate-lanceolate to lanceolate, strongly falcate, usually 8–18, up to 9 × 3 cm., apex long-acuminate, base very asymmetric, margin coarsely serrate, glabrous. Inflorescence an axillary panicle up to 35 cm. long. Petals white. Staminal tube pale cream. Drupe ellipsoid, 1.5–1.8 cm. long, yellow. Fig. 6.

UGANDA. Mengo District; Entebbe Botanic Garden, cult. Sept. 1962, *Styles* 21!
KENYA. Kilifi District: Malindi, naturalized, Nov. 1962, *Greenway* 10847!; Lamu District: Lamu town, Oct. 1980, *Hume* 10!
TANZANIA. Lushoto District: Mombo Arboretum, Feb. 1971, *Ruffo* 381!; Tabora District: Uruma [Urumwa] experimental plot 669, Oct. 1970, *Ruffo* 426!; Uzaramo District: Dar es Salaam, Jan. 1977, *Wingfield* 3761! & 3 Sept. 1946, *Wigg* in *F.H.* 1548!; Zanzibar I., Tomondo, July 1930, *J.H. Vaughan* 1410!; Pemba I., Weti, *W.E. Taylor* 353!
DISTR. U 4; K 7; T 3, 4, 6; Z; P; native of India and Burma and now widely planted in the drier parts of Africa and naturalized locally
HAB. Frequently grown in and around villages, and locally in plantations (but with little success). In NE. Kenya it is the most frequently planted street tree and has escaped and regenerated plentifully; 0–1400 m.

SYN. *Melia azadirachta* L., Sp. Pl.: 385 (1753): Type as for *Azadirachta indica*

NOTE. Both in its native lands and elsewhere, *Azadirachta indica*, the Neem, is one of the most useful of all trees. The wood is hard, red, durable and attractive, and is used for housebuilding and furniture and as fuel. All parts of the plant are medicinal. A powerful insect-antifeedant, the limonoid azadirachtin, occurs in this species and is used to control various types of insect larvae including locusts (see F. White, loc. cit., 1986).

4. PSEUDOBERSAMA

Verdc. in J.L.S. 55: 504, t. 1 (1956); Pennington & Styles in Blumea 22: 469 (1975)

Trees. Leaves imparipinnate, alternate; leaflets entire. Flowers unisexual (plant dioecious) in compound cymes. Calyx cupuliform with 5 teeth. Petals 5, free, imbricate. Stamens 11–12; filaments connate at the base from $\frac{1}{3}$–$\frac{2}{3}$ of their length; anthers puberulous. Disk annular, partly fused to the base of the ovary or pistillode. Pistillode conical at base, cylindric, distally hirsute, with 3 locules, sometimes with vestigial ovules. Staminodes 12, filaments connate; antherodes hirsute, indehiscent, not producing pollen. Ovary ovoid-globose, densely hirsute, narrowed into the short style; style-head scarcely broader than the style, obscurely 4–5-lobed; locules 5, each with 2 ovules. Fruit a loculicidal capsule with 4–5 thick woody valves bearing branched antler-like appendages outside. Seeds very small in relation to the size of the capsule, partly covered by a bright red aril.

One species on the eastern side of Africa.

P. mossambicensis *(Sim) Verdc.* in J.L.S. 55: 504, t. 1 (1956); K.T.S.: 270 (1961); F. White & Styles in F.Z. 2: 305, t. 60 (1963) & in F.S.A. 18(3): 57, fig. 18 (1986); F. White in Bothalia 16: 161 (1986). Type: Mozambique, without locality, *Sim* 5204 (not located); fig. 23 in Sim., For. Fl. Port. E. Afr. (1909), lectotype of White (1986)

FIG. 7. *PSEUDOBERSAMA MOSSAMBICENSIS*—**1**, fruiting twig, × ⅔; **1a**, half ♂ flower, vertical section, × 5; **1b**, capsule, × ⅔; **1c**, seed, × ⅕. 1, from *Boococh* 16; 1a, after Pennington & Styles; 1b, 1c, from *Moggridge* 430. Drawn by Julia Loken.

Small or medium-sized evergreen tree up to 20 m. tall, sometimes flowering as a shrub 2–4 m. tall. Leaves imparipinnate; petiole and rhachis up to 30 cm. long, minutely puberulous towards base; leaflets 9–17, alternate or subopposite, elliptic or oblong-elliptic, up to 15 × 6 cm., usually smaller, apex shortly and bluntly acuminate, base cuneate, asymmetric, the proximal leaflets smaller and proportionally broader, young leaflets densely pubescent, glabrous above when mature, lower surface with conspicuous tufts of hairs in nerve-axils and prominent open reticulate venation. Flowers white, 3–12, in lax to subcapitate compound cymose inflorescences, each part of the inflorescence bearing a simple 3-flowered cyme; peduncle 1–6 cm. long; bracts subulate, 0.1–0.3 cm. long, puberulous; pedicels 0.1 cm. long. Calyx ± 0.3 cm. long, lobed to ± the middle; lobes deltate. Petals 0.5–0.6 cm. long, glabrous except for minute papillae. Filaments ± 0.35 cm. long, united for ⅕–⅔ of their length, densely hairy towards the apex; anthers 0.15 cm. long, hairy. Pistillode 0.3 cm. long. Antherodes 0.1 cm. long. Ovary 0.3 × 0.3 cm.; style 0.15 cm. long. Capsule 3–4.5 cm. in diameter, very woody, red, densely covered with lobed antler-shaped appendages 0.7 cm. long, dehiscing by (4–)5(–6) valves, which remain connate at the base; valves 0.4–0.6 cm. thick; stipe ± 0.5 cm. long; seeds 2 per locule, 0.7 × 0.5 cm., purple-black; aril bright red, confined to adaxial part of seed, and forming a cushion at the apex and base. Fig. 7.

KENYA.   Kwale District: Shimba Hills, Mandakara Forest, Dec. 1975, *White* 11319!; Kilifi District: 8 km. NE. of Kilifi, Feb. 1961, *Greenway* 9814! & Sokoke Forest, Apr. 1954, *Trump* 141!
TANZANIA.   Uzaramo District: Pugu Forest Reserve, June 1954, *Semsei* 1762!; Ulanga District: Magombera Forest Reserve, Nov. 1961, *Semsei* 3371!; Kilwa District: Pungatini, Apr. 1912, *Braun* 3688!
DISTR.   K 7; T 3, 6, 8; Mozambique and South Africa (northern Natal)
HAB.   In understorey and at edges of moist lowland forest; 60–300 m.

SYN.   *Bersama mossambicensis* Sim, For. Fl. Port. E. Afr.: 34 (1909); T.S.K., ed. 2: 110 (1936); T.T.C.L.: 324 (1949)

NOTE.   The poles are used in hut building (*MacNaughton* 119)

5. TRICHILIA

P. Browne, Hist. Jamaica: 278 (1756); L., Syst. Nat., ed. 10: 1020 (1759); J.J. de Wilde, Revision of the species of *Trichilia* on the African Continent, in Meded. Landbouwhogeschool Wageningen 68-2: 1–207 (1968); Pennington & Styles in Blumea 22: 467 (1975), *nom. conserv.*

Trees, often of large size, or rarely shrublets. Leaves imparipinnate, rarely 3-foliolate or 1-foliolate. Flowers unisexual (plants dioecious) in cymes or cymose panicles. Calyx cupuliform with 5 minute teeth, or deeply lobed with 5 deltate or circular lobes. Petals 5, free, imbricate. Filaments 10, completely united to form a staminal tube bearing the anthers on its ± entire margin or united only in the lower half and then anthers inserted between a pair of deltate appendages; anthers glabrous. Disk completely fused to the base of the staminal tube or cup-shaped and free from it. Ovary small, with 2–4 locules, each with 2 collateral or superposed ovules; style elongate, much longer than the ovary, distally expanded to form a capitate or annular style-head surmounted by a lobed stigmatic area. Pistillode similar to the gynoecium but narrower, usually with vestigial ovules. Fruit a loculicidal capsule with (2–)3(–4) ± leathery rather thin valves. Seeds large, each partly or almost completely covered by a bright red aril, or the aril apparently absent and then the testa fleshy.

A pantropical genus with about 70 species in America and 18 in Africa.

Bole deeply fluted for most of length; leaflets rarely more than
    7, venation indistinct; filaments united almost to apex,
    without appendages   .   .   .   .   1. *T. prieuriana*
Bole not fluted; leaflets usually more than 7, venation clearly
    visible; filaments free in upper half, each with a pair of
    deltate appendages:
  Petals up to 0.7 cm. long; disk free from staminal tube;
    capsule glabrous:

Petioles with knife-like edges at the base; petals 0.4–0.7 cm.
  long; capsule transversely wrinkled   . . . . .   2. *T. rubescens*
Petioles without knife-like edges; petals up to 0.3 cm. long;
  capsule verruculose   . . . .   3. *T. lovettii*
Petals more than 0.9 cm. long; disk united with staminal
  tube; fruit hairy:
    Leaflets broadest near the base, gradually tapering to an
      attenuate-acuminate apex; petiolules of lower leaflets
      slender, 1–1.5 cm. long   . . . .   4. *T. martineaui*
    Leaflets broadest near the middle or in upper half; apex
      rounded to shortly and suddenly acuminate:
        Capsule without a stipe or with an indistinct stipe up to
          0.3 cm. long   . . . .   5. *T. dregeana*
        Capsule sharply differentiated from a 0.5–1 cm. long
          stipe   . . . .   6. *T. emetica*

1. **T. prieuriana** *A. Juss.* in Bull. Sci. Nat. 23: 238 (Nov. 1830) & in Mém. Mus. Nat. Hist. Nat., Paris 19: 236 & 276 (1830) as '*prieureana*'; F.T.A. 1: 334 (1868); C. DC. in A. & C. DC., Monogr. Phan. 1: 678 (1878); V.E. 3(1): 823 (1915); Vermoesen in Rev. Zool. Afr. 10, Suppl. Bot.: 46 (1922); Pellegrin in Not. Syst. 9: 17, fig. 1C (1940); C.F.A. 1(2): 312 (1951); I.T.U., ed. 2: 197 (1952); Staner & Gilbert in F.C.B. 7: 164 (1958); F.W.T.A., ed. 2, 1: 704 (1958); F. White & Styles in F.Z. 2: 304, fig. 58C (1963); J.J. de Wilde in Meded. Landbouwhogeschool Wageningen 68-2: 130, fig. 12A–C (1968); Styles in E. Afr. Agric. Journ. 39: 419 (1974); Styles & F. White in Fl. Eth. 3: 483, fig. 124.3/1–3 (1990). Lectotype chosen by de Wilde (1968): Senegal, Casamance R., *Perrottet* (P, lecto.!)

Evergreen tree usually 10–15(–25) m. tall, sometimes flowering as a shrub; bole fluted; bark rough, peeling easily. Leaves up to 25 cm. long; leaflets (5–)7–9(–11), usually drying a dull grey-green, elliptic or oblanceolate-elliptic, up to 18 × 6.5 cm., apex shortly and bluntly acuminate, base cuneate, lower surface glabrous with indistinct venation. Inflorescence a much-branched congested panicle, usually less than 7 cm. long, in axils of leaves or of fallen leaves. Calyx 0.2–0.3 cm. long, puberulous, lobed to the middle, lobes deltate. Petals 0.4–0.8 cm. long, puberulous. Staminal tube 0.5–0.7 cm. long. Capsule globose, ± 2 × 2 cm., glabrous, slightly sulcate, not or scarcely wrinkled; stipe absent. Seeds very dark brown; aril orange-red. Fig. 8/1–3.

UGANDA. Lango District: Badyang Forest Reserve, Apr. 1945, *Greenway & Eggeling* 7342!; Teso District: Serere, Feb. 1933, *Chandler* 1099!; Mengo District: Mabira Forest, Nov. 1962, *Styles* 219!
TANZANIA. Kigoma District: Gombe National Park, Kahama valley, *Clutton-Brock* 315!; Mpanda District; Mahali Mts., Lwengele, July 1978, *Uehara* 525!
DISTR. U 1–4; T 4; in Guineo-Congolian rain-forest from Senegal eastwards and with outliers in Ethiopia and Zambia
HAB. Lowland rain-forest and riparian forest; 915–1220 m.

SYN. *T. euryphylla* Burtt Davy & Bolton, Check-Lists For. Trees & Shrubs Brit. Emp. 1, Uganda: 65 (1935), *nom. nud.*
  T. prieuriana A. Juss. subsp. *vermoesenii* J.J. de Wilde in Meded. Landbouwhogeschool Wageningen 68-2: 139, fig. 12B (1968). Type: Zaire, Temvo, *Vermoesen* 1829 (BR, holo.!, K, iso.!)
  T. prieuriana A. Juss. subsp. *orientalis* J.J. de Wilde in Meded. Landbouwhogeschool Wageningen 68-2: 146, fig. 12C (1968). Type: Uganda, Acholi District, Paimol, *Dawkins* 310 (K, holo.!)

NOTE. The three subspecies recognized by de Wilde are based on minor and not very constant differences of floral structure and indumentum. Subsp. *vermoesenii* occurs in the south and west of Uganda and subsp. *orientalis* in the north and east of Uganda and in Tanzania.
  T. prieuriana was formerly regarded as a weed tree by the Uganda Forest Department and was poisoned. Its timber, which splits easily, has not been used in the timber trade or locally.

2. **T. rubescens** *Oliv.* in F.T.A. 1: 336 (1868); C. DC. in A. & C. DC., Monogr. Phan. 1: 708 (1878); Harms in Z.A.E.: 433 (1912); V.E. 3(1): 822 (1915); Vermoesen in Rev. Zool. Afr. 10, Suppl. Bot.: 50 (1922); Pellegrin in Not. Syst. 9: 20, fig. 1D (1940); I.T.U., ed. 2: 197 (1952); Staner & Gilbert in F.C.B. 7: 169 (1958); F.W.T.A., ed. 2, 1: 704 (1958); Verdc. in K.B. 14: 346 (1960); J.J. de Wilde in Meded. Landbouwhogeschool Wageningen 68-2: 161, fig. 15 (1968); Styles in E. Afr. Agric. Journ. 39: 420 (1974). Lectotype chosen by de Wilde (1968): Cameroon, Ambas Bay, *Mann* 20 (K, lecto.!)

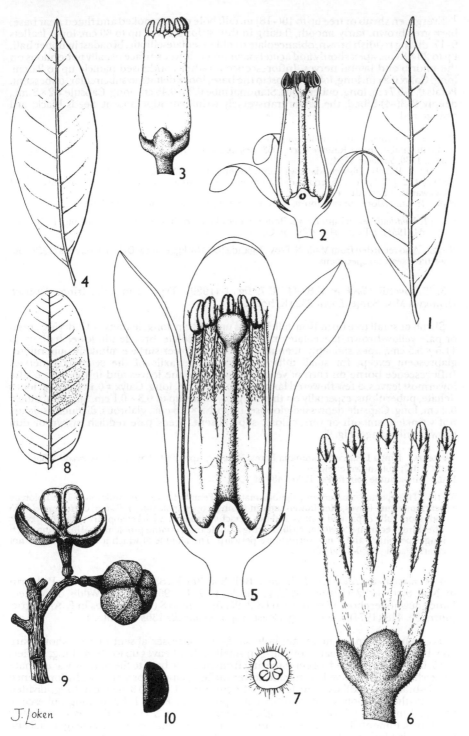

FIG. 8. *TRICHILIA PRIEURIANA*—1, leaflet, × ²/₅; 2, half ♂ flower, vertical section, × 3¹/₁₀; 3, calyx and staminal tube, × 3¹/₁₀. *T. DREGEANA*—4, leaflet, × ²/₅; 5, half ♀ flower, × 3¹/₁₀; 6, calyx and staminal tube, × 3¹/₁₀; 7, transverse section of ovary, × 3¹/₁₀. *T. EMETICA*—8, leaflet, × ²/₅; 9, fruits, × ²/₅; 10, seed, × ²/₅. 1, from *Styles* 226; 2,3 from *Styles* 219; 4, from *Styles* 206; 5–7, from *Styles* 333; 8, from *Styles* 254; 9,10, from *Bond* s.n. Drawn by Julia Loken.

Evergreen shrub or tree up to 10(–18) m. tall; bole often crooked and fluted near base; bark grey-brown, fairly smooth, flaking in thin strips. Leaves up to 30 cm. long; leaflets 9–17, drying reddish brown, oblanceolate or oblanceolate-elliptic. broadest in upper half, up to 20 × 8 cm., apex shortly and acutely acuminate, lower surface usually puberulous on the midrib and lateral nerves. Inflorescence a rather condensed panicle up to 24 cm. long. Calyx 0.2 cm. long, lobed almost to the base; lobes deltate to subcircular, pubescent. Petals 0.4–0.7 cm. long, puberulous. Staminal tube 0.25–0.45 cm. long. Capsule ± 2 × 2 cm., deeply 2–3(–4)-lobed, the valves transversely wrinkled; stipe absent. Seeds black; aril orange-red.

UGANDA. Bunyoro District: Budongo Forest Reserve, Jan. 1963, *Styles* 330 & 330a!; Masaka District: Malabigambo Forest Reserve, Oct. 1953, *Drummond & Hemsley* 4583!; Mengo District: Kyagwe, Sept. 1930, *Snowden* 1753!
TANZANIA. Bukoba District: Kiamawa Forest Reserve, Oct. 1935, *Gillman* 407!
DISTR. U 2, 4; T 1; from Nigeria and Cameroon southwards to near the mouth of the Zaire R. and eastwards to East Africa
HAB. In understorey of rain-forest, both on dry ridges and in swamps; 900–1500 m.

SYN. [*T. heudelottii* sensu Burtt Davy & Bolton, Check-Lists For. Trees & Shrubs Brit. Emp. 1, Uganda: 64 (1935); I.T.U.: 104 (1940), *non* Oliv.]

NOTE. Also recorded from Madi, N. Prov., Uganda (U 1) by Eggeling & Dale, I.T.U., ed. 2 (1952), but we have seen no specimens.

3. **T. lovettii** *Cheek* in K.B. 44: 457, fig. 1 (1989). Type: Tanzania, Iringa District, Uzungwa Mts., Sanje, *Lovett* 232 (K, holo.!)

Shrub or small tree up to 10 m. tall. Leaves up to 18 cm. long; leaflets 7–11, drying green or pale yellow-brown, lanceolate or oblong-oblanceolate, broadest in lower half, up to 11.5 × 3.5 cm., apex suddenly attenuate-acuminate, lower surface minutely puberulous, glabrescent except for small tufts of hairs in the axils of the secondary nerves. Inflorescence borne on current year's shoots below the leaves and in the axils of the lowermost leaves, a few-flowered lax cyme up to 3.5 cm. long. Calyx ± 0.1 cm. long; lobes deltate, puberulous, especially on the margins. Petals up to 0.3 × 0.1 cm. Staminal tube ± 0.2 cm. long. Capsule depressed-globose, up to 1.2 × 1.8 cm., glabrous, distinctly sulcate, not wrinkled, smooth or verruculose; stipe absent. Seeds pale reddish brown or dull brown; aril orange. Fig. 9.

TANZANIA. Iringa District: Mwanihana Forest Reserve, Apr. 1981, *Rodgers & Homewood* 1035!
DISTR. T 7; not known elsewhere
HAB. Mid-altitude rain-forest; 1220–1480 m.

NOTE. This species is based on exiguous material. Female flowers are unknown. The holotype bears immature leaves, a few flower-buds and a single open male flower. *Rodgers & Homewood* 1035 with more fully expanded leaflets and two detached fruits in a packet appears to be conspecific The third specimen (*Lovett et al.* 8546, bearing a single fruit) is more problematical in having much longer petiolules and a non-verruculose pericarp. The altitude at which it was collected is not clearly stated ('1250–1700 m').

4. **T. martineaui** *Aubrév. & Pellegrin* in Bull. Soc. Bot. Fr. 83: 491, fig. 2A (1936); Pellegrin in Not. Syst. 9: 22 (1940); F.W.T.A., ed. 2, 1: 704 (1958); J.J. de Wilde in Meded. Landbouwhogeschool Wageningen 68-2: 96, fig. 8, map 8 (1968); Styles in E. Afr. Agric. Journ. 39: 418 (1974). Type: Ivory Coast, Yapo, *Aubréville* 1365 (P, holo.!)

Tree up to 35 m. or more tall; bole straight; buttresses absent or very short; bark grey-brown or black, rough, sloughing in small scales. Leaves up to 26 cm. long; leaflets 9–15, lanceolate, up to 14 × 4 cm., tapering from near the base to the narrowly acuminate apex which is often curved to one side, lower surface glabrous or almost so. Inflorescence a little-branched axillary panicle up to 12 cm. long. Calyx 0.3–0.35 cm. long, divided almost to the base; lobes ovate to subcircular, pubescent. Petals 1–1.2 cm. long, pubescent. Staminal tube 0.5–0.7 cm. long. Capsule subglobose, ± 3 cm. in diameter, shallowly 2–3-lobed, tomentellous; valves with indistinct longitudinal wrinkles; stipe up to 1.5 cm. long. Seeds black, almost completely concealed by the orange-red aril.

UGANDA. Bunyoro District: Budongo Forest, May 1941, *Eggeling* 4375!; Mengo District: Mabira Forest Reserve, Sept. 1947, *Kigundu* 80/62!; & Kifu Forest Reserve, May 1972, *Stuart Smith* 259!

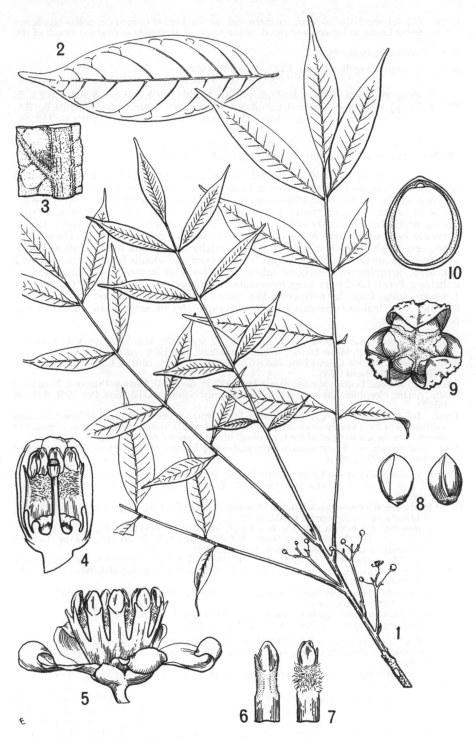

FIG. 9. *TRICHILIA LOVETTII*—1, habit, × ⅖; 2, mature leaflet, × ⅖; 3, detail of domatium, × 12; 4, median section of unopened ♂ flower × 10; 5, whole ♂ flower, × 10; 6, stamen, external view, × 12; 7, stamen, internal view, × 12; 8, seeds, ventral view, × 1½; 9, verrucate fruit, × 1½; 10, seed, × 3. 1, 3–7 from *Lovett* 232; 2, 10, from *Lovett et al.* 854b; 8, 9, from *Rodgers & Homewood* 1035. Drawn by Eleanor Catherine.

DISTR.   U 2, 4; towards the northern, southern and eastern limits of Guineo-Congolian rain-forest from Sierra Leone to Uganda and (south of the equator) westwards to near the mouth of the Zaire R.
HAB.   Rain-forest; 1000–1200 m.
SYN.   *T. sp. 1* sensu Eggeling & Dale, I.T.U., ed. 2: 199 (1952)

5. **T. dregeana** *Sond.* in Harv. & Sond., Fl. Cap. 1: 246 (1860); F. White & Styles in F.Z. 2: 298 (1963); J.J. de Wilde in Meded. Landbouwhogeschool Wageningen 68-2: 28, fig. 3A, map 3 (1968); Styles in E. Afr. Agric. Journ. 39: 418 (1974); F. White in Bothalia 16: 158, fig. 6/1 (1986); F. White & Styles in F.S.A. 18(3): 55, fig. 17/1 (1986); Styles & F. White in Fl. Eth. 3: 483, fig. 124.3/4–7 (1990). Lectotype chosen by de Wilde (1968): South Africa, Natal, Durban [Port Natal], *Gueinzius* (TCD, lecto., G!, K!, S, isolecto.)

Evergreen tree up to 20–40 m. tall; bole sometimes slightly buttressed; bark grey and smooth, with shallow fissures. Leaves up to 26 cm. long; leaflets usually 7–11, obovate to oblanceolate or oblanceolate-elliptic, nearly always broadest in upper half, up to 21 × 8.5 cm., apex of lateral leaflets shortly and acutely acuminate or bluntly subacuminate (with the tip itself often shallowly notched), nearly always showing a hollow curve; lateral nerves in 7–12, usually widely spaced pairs; lower surface glabrous or with a few strigose hairs, very rarely sparsely to densely pilose, usually drying dark brown. Inflorescence sometimes many-flowered in ♂, always few-flowered and usually lax in ♀. Calyx 0.5–0.7 cm. long, strigulose-tomentellous, lobed to half-way or more; lobes subcircular or imbricate. Petals 1.3–2.6 cm. long, tomentellous on both surfaces. Staminal tube usually 1–1.6 cm. long. Capsule (unopened, but mature) 3–5 cm. in diameter, occurring 1–3 together in the leaf-axils. Seeds black, almost completely concealed by the scarlet aril. Fig. 8/4–7, p. 31.

UGANDA.   Bunyoro District: Budongo Forest, Jan. 1963, *Styles* 333!; Ankole District: Kalinzu Forest, Oct. 1962, *Styles* 162!; Mbale District: 16 km. E. of Mbale, Jan. 1963, *Styles* 321!
KENYA.   S. Nyeri District: Chania Falls, *Faden* 67/45!; N. Kavirondo District: Kakamega Forest, Jan. 1968, *Perdue & Kibuwa* 9452!
TANZANIA.   Bukoba District: Minziro Forest, Jan. 1959, *Procter* 1124!; Lushoto District: E. Usambara Mts., Amani, Nov. 1935, *Greenway* 4157!; Ufipa District: Namwele Coal Mine, Dec. 1961, *Richards* 15802!
DISTR.   U 1–4; K 4, 5; T 1–4, 6, 7; widespread in forest from Guinée to Ethiopia and East Africa and southwards to the Cape Province in South Africa and with a westward extension to Angola, but absent from the greater part of the rain-forests of the Guineo-Congolian region
HAB.   Mid-altitude rain-forest; riparian forest and swamp forest; sometimes planted in arboreta and gardens; (500–)775–1525 m.
SYN.   *T. stuhlmannii* Harms in E.J. 23: 162 (1896); Engler in V.E. 3(1); 820 (1915); T.T.C.L.: 319 (1949).
   Types: Tanzania, Bukoba, *Stuhlmann* 1136 (B, holo.†) & *Gillman* 311 (K, neo.! of de Wilde (1968))
   T. chirindensis Swynnerton & Bak.f. in J.L.S. 40: 39 (1911); T.T.C.L.: 318 (1949): Type: Zimbabwe, Chirinda Forest, *Swynnerton* 1 (BM, holo.!, K, iso.!)
   T. splendida A. Chev. in Bull. Soc. Bot. Fr. 58, Mém. 8: 147 (1912); I.T.U., ed. 2: 197 (1952); F.W.T.A., ed. 2, 1: 705 (1958); Staner & Gilbert in F.C.B. 7: 165 (1958). Type: Guinée, Kissidougou, *Chevalier* 20708 (P, holo.!, BR, K, iso.!)
   T. schliebenii Harms in N.B.G.B. 11: 1070 (1934); T.T.C.L.: 319 (1949). Type: Tanzania, Morogoro District, Uluguru Mts., *Schlieben* 3636 (B, holo.†, G, lecto.! of de Wilde (1968), BM, BR, P, Z, isolecto.!)
   T. redacta Burtt Davy & Bolton, Check-lists For. Trees & Shrubs, Brit. Emp. 1, Uganda: 64 (1935); I.T.U.: 105 (1940), *nom nud.*
   [*T. megalantha* sensu Eggeling & Dale, I.T.U., ed. 2: 197 (1952), *non* Harms]
   T. sp. 2 sensu Eggeling & Dale, I.T.U., ed. 2: 199 (1952)
NOTE.   For economic uses see F.Z. 2: 299 (1963). In Uganda the excellent reddish brown timber which resembles other African mahoganies is useful for furniture. Large quantities however are not available.
   The relationships of *T. dregeana* and *T. emetica* are discussed under the latter.

6. **T. emetica** *Vahl*, Symb. Bot. 1: 31 (1790); DC., Prodr. 1: 622 (1824); F.T.A. 1: 335 (1868); C. DC. in A. & C. DC., Monogr. Phan. 1: 660 (1878); Gürke in P.O.A. C: 231 (1895); Bak. in J.L.S. 37: 133 (1905); V.E. 3(1); 821, fig. 387 & 388L–S (1915); Vermoesen in Rev. Zool. Afr. 10, Suppl. Bot.: 34 (1922); T.T.C.L.: 318 (1949); C.F.A. 1: 314 (1951); I.T.U., ed. 2: 196, fig. 46 (1952); F. White & Styles in F.Z. 2: 299, t. 58B1–4 (1963); J.J. de Wilde in Meded. Landbouwhogeschool Wageningen 68-2: 50, fig. 4A, B (1968) excl. syn. *T. grotei;* Styles in F. Afr. Agric. Journ. 39: 418 (1974); F. White in Bothalia 16: 158, fig. 6/2 (1986); F. White &

Styles in F.S.A. 18(3): 57, fig. 17/2 (1986); Styles & F. White in Fl. Eth. 3: 483, fig. 124.3/8–10 (1990). Type: Yemen, Hadie Mts., *Forsskål* 478 (C, holo., BM, iso.!)

Evergreen or semi-evergreen tree, usually 8–20(–25) m. tall; crown very dense, widespreading in open; foliage very dark green; bark dark grey or dark brown, rough or smooth. Leaves up to 28 cm. long; leaflets usually 9–11, elliptic or oblong-elliptic, up to 15 × 5 cm., nearly always broadest near the middle and with the apex of lateral leaflets rounded, emarginate or broadly acute without a hollow curve; lateral nerves usually in 11–18 closely set pairs; lower surface sparsely to densely puberulous with short, weak curly or flexuous hairs, usually drying olive-green or pale yellow-brown. Inflorescence usually condensed and many-flowered. Calyx 0.35–0.5 cm. long, tomentellous, lobed almost to the base; lobes suborbicular, imbricate. Petals (0.7–)1–1.6 cm. long. Staminal tube usually 0.8–1.1 cm. long. Capsule (unopened but mature) 1.8–2.5 cm. in diameter, usually crowded at the ends of the branchlets. Seeds as in *T. dregeana*. Fig. 8/8–10, p. 31.

UGANDA. W. Nile District: Koboko, Feb. 1934, *Eggeling* 1498!; Acholi District: Abera Forest Station, Dec. 1962, *Styles* 254!; Mbale District: Muyembe, Jan. 1963, *Styles* 322!
KENYA. Northern Frontier Province: Katilia, Kerio R., Aug. 1968, *Mwangangi & Gwynne* 1186!; Machakos District; Kibwezi, Apr. 1972, *Gillett* 19659!; Kwale District: Diani Forest, Dec. 1975, *White* 11399!
TANZANIA. Mwanza District: Saanane I., Feb. 1965, *Carmichael* 1188!; Rufiji District: Mafia I., Juani I., Sept. 1937, *Greenway* 5294!; Rungwe District: Kyela, *Harris* 2531!; Zanzibar I., Chwaka, Oct. 1937, *J.H. Vaughan* 2004!
DISTR. U 1–3; K 1, 3–7; T 1–8; Z; widespread in Africa but absent from the Guineo-Congolian region; from Senegal to the Red Sea and southwards through East and Central Africa to the Caprivi Strip, Botswana and South Africa (Natal); also in the Arabian Peninsula
HAB. In coastal forest, drier types of riparian forest and riparian woodland; more rarely in rocky outcrops or in wooded grassland; 10–1300 m.

SYN. *T. somalensis* Chiov. in Result. Sci. Miss. Stef.-Paoli, Coll. Bot. 1: 50 (1916). Type: Somalia, Hàcacca, *Paoli* 515 (FT, holo.!)
   T. grotei Harms in N.B.G.B. 7: 230 (1917); T.T.C.L.: 319 (1949). Type: Tanzania, E. Usambara Mts., Amani, near Kihuhwi [Kiuhui], *Grote* 3774 (B, holo.†, K, lecto.! of de Wilde (1968), EA, isolecto.!)
   T. roka Chiov. in Bol. Soc. Bot. Ital. 1923: 115 (1923) & in Fl. Somala 2: 131 (1932); Staner & Gilbert in F.C.B. 7: 163 (1958); F.W.T.A., ed. 2, 1: 705 (1958); K.T.S.: 270 (1961), *nom. illegit.* Type as for *T. emetica*
   T. jubensis Chiov., Fl. Somala 2: 129, fig. 88 (1932). Type: Somalia, Juba [Guiba] R. near Alessandria, Touata I., *Tozzi* 326 (FT, holo.!, K, iso.!)
   T. sp. nov. sensu Dale & Greenway, K.T.S.: 272 (1961)

NOTE. Throughout its range, *T. emetica* has many uses (see, for example, F.Z. 2: 300 (1963)). In South Turkana the trunk is used for making water-carrying vessels (*Mwangangi & Gwynne* 1186). The use of *T. emetica* as an emetic and for other medicinal purposes has long been known, occasionally resulting in death. This happens, apparently, only if the dose is too high or if the patient, for other reasons of ill-health, is not suitable for such a powerful remedy (Govt. Chemist, Kenya in E.A.H. 14878).

   The reasons for rejecting the name *T. roka* Chiov. are discussed elsewhere (F. White in Bothalia 16: 157 (1986)), where the differences between *T. dregeana* and *T. emetica* in Southern Africa are analysed in detail. A similar comparison based on East African material has shown that the same characters apply there, except that in the coastal forests of Kenya and Tanzania the leaflets of *T. emetica* are often more tapered to the apex and base and confusion with *T. dregeana* is possible, unless a careful comparison is made. The flowers and fruits of the coastal plant, however, are typical of *T. emetica*.

   de Wilde (1968) recognizes two subspecies of *T. emetica*. Subsp. *suberosa* J.J. de Wilde (p. 67) extends from Senegal to Uganda, where it overlaps with the typical subspecies. Its branchlets are covered with corky bark and its characteristic habitat is woodland or wooded grassland, where it can withstand the effects of fire. Other differences are rather slight, and the bark character also occurs sporadically in the range of subspecies *emetica*, e.g. *Harris* 2431 from Tanzania. Pending further investigation the status and distribution of subsp. *suberosa* in East Africa is uncertain.

Doubtful species

T. subcordata *Gürke* in P.O.A. C: 232 (1895); Harms in E.J. 23: 163 (1896); V.E. 3(1): 821 (1915); T.T.C.L.: 319 (1949). Type: Tanzania, Tanga District, Amboni, *Holst* 2723 (B, holo.†)

NOTE. de Wild in Meded. Landbouwhogeschool Wageningen, 68-2: 198 (1968) doubts that this species is a *Trichilia*.

FIG. 10. *LEPIDOTRICHILIA VOLKENSII*—1, flowering shoot, × ½; 2, half flower, vertical section, × 6¾; 3, infructescence, × ½; 4, galled fruits, × ½; 5, stellate hair, × 225. *EKEBERGIA CAPENSIS*—6, flowering shoot, × ½; 7, half ♀ flower, vertical section, × 6¾; 8, infructescence, × ½; 9, fruit, transverse section, × ½. 1,5, from *de Wilde* 4405; 2, from *de Wilde* 7034; 3, from *Styles* 180; 4, from *de Wilde* 5288; 6, from *Bos* 9720; 7, from *White* 8146; 8,9, from *de Wilde* 5992. Drawn by Rosemary Wise.

Invalidly published species

T. holtzii Harms in E. & P. Pf., ed. 2, 19B(1): 110 (1940), is without a latin description; T.T.C.L.: 319 (1949). See de Wilde, loc. cit. 203 (1968).

6. LEPIDOTRICHILIA

J. Leroy in Comptes Rendus Acad. Sci., Paris 247: 1025 (1958) & in Journ. Agric. Trop. Bot. Appl. 5: 673 (1958); Pennington & Styles in Blumea 22: 473 (1975)

Trees. Leaves imparipinnate; leaflets entire, stellate-puberulous. Flowers ♂, borne in large cymose panicles. Calyx cupuliform, with 5 minute teeth. Petals 5, free, much longer than the calyx in bud, induplicate-valvate. Filaments 10, fused in the lower half to form a staminal tube; anthers inserted between a pair of deltate-acuminate appendages which slightly exceed them in length. Disk absent, but base of staminal tube apparently slightly modified for nectar secretion. Ovary (in Flora area) with 2–4 locules, each locule with 1 ovule; style-head capitate, surmounted by 2–4 erect stigmatic lobes. Fruit a berry.

1 species in Africa and 3 in Madagascar.

L. volkensii (*Gürke*) *J. Leroy* in F. White & Styles in F.Z. 2: 305, t. 58B1–4 (1963); Styles in E. Afr. Agric. Journ. 39: 414 (1974); Styles & F. White in Fl. Eth. 3:485, fig. 124.6/1–5 (1990). Types: Tanzania, Kilimanjaro, Marangu, near Moonjo [Mondjo] Stream, *Volkens* 1243 (B, syn.†, K, lecto.! here designated) & Marangu, *Volkens* 1269 (B, syn.†)

Small or medium-sized evergreen tree 5–20 m. tall, sometimes flowering as a shrub; bole fluted; bark smooth, grey. Leaves imparipinnate; petiole and rhachis up to 30 cm. long, densely stellate-pubescent; leaflets opposite or alternate, 7–11, very variable in shape and size, up to 16.5 × 7.5 cm., usually much smaller, distal leaflets ± elliptic, apex acute to shortly acuminate, base markedly asymmetric, upper surface drying pale yellow-green, lower surface puberulous with small stellate hairs and minute red and black glands; petiolules up to 1 cm. long. Flowers creamy white, becoming yellow with age, fragrant, in contracted cymose panicles in leaf-axils; peduncles up to 12 cm. long; pedicels 0.1–0.25 cm. long. Calyx 0.2 cm. long, cupuliform, with minute teeth, scurfy-stellate-pubescent. Petals 0.4–0.5 cm. long, densely puberulous outside. Filaments 0.3–0.4 cm. long, glabrous or puberulous at the base outside, sparsely villous in the upper half inside. Style 0.1–0.2 cm. long. Berry depressed-globose, small, up to 1.5 × 1 cm., shallowly sulcate, encrusted with stellate scales, 2–3-locular, with 1 dark brown or black exarillate seed in each locule. Fig. 10/1–5.

UGANDA. Acholi District: SE. Imatong Mts., Aringa headwaters, Apr. 1945, *Greenway & Hummel* 7301!; Ankole District: Kalinzu Crown Forest Reserve, Oct. 1962, *Styles* 180!; Mbale District: Elgon Crown Forest Reserve, Nkonkonjeru Peninsula, Jan. 1963, *Styles* 320!
KENYA. Baringo District: Katimok Forest, Oct. 1930, *Dale* in *F.D.* 2431!; N. Kavirondo District: Kakamega Forest, Apr. 1973, *Hansen* 898!; Teita Hills, Yale Peak, Sept. 1953, *Drummond & Hemsley* 4324!
TANZANIA. Arusha District: Mt. Meru, Ngongongare Forest, Jan. 1954, *Hughes* 189!; Njombe District: Lisitu–Lugalawa, Sept. 1970, *Thulin & Mhoro* 1119!; Songea District: Liwiri–Kiteza Forest Reserve, Oct. 1956, *Semsei* 2524!
DISTR. U 1–4; K 1, 3–7; T 2–4, 6–8; Ethiopia, Sudan, E. Zaire, Rwanda, Burundi, Malawi
HAB. In montane and mid-altitude forest; 1550–2600 m.

SYN. *Trichilia volkensii* Gürke in E.J. 19, Beibl. 47: 33 (1894) & in P.O.A. C: 232 (1895); Harms in Z.A.E.: 435 (1912); V.E. 3(1): 822, fig. 388T–V (1915); T.T.C.L.: 319 (1949); I.T.U., ed. 2: 198 (1952); Staner & Gilbert in F.C.B. 7: 160 (1958); K.T.S.: 271 (1961)
   Commiphora kilimandscharica Engler, P.O.A. C: 230 (1895). Type: Tanzania, Kilimanjaro, *Volkens* 1849 (B, holo.†)
   Trichilia volkensii Gürke var. *genuina* Pichi Serm. in Webbia 7: 334 (1950), *nom. superfl.*

7. EKEBERGIA

Sparrm. in Sv. Vet. Akad. Handl. 40: 282, t. 9 (1779)

Trees or shrubs. Indumentum of simple hairs. Leaves imparipinnate. Flowers unisexual (plant dioecious), in contracted panicles. Calyx (4–)5(–6)-lobed in upper half. Petals (4–)5, free, imbricate. Staminal tube cup-shaped, filaments completely fused or almost so, without appendages; anthers usually 10, hairy or glabrous, inserted on the rim of the staminal tube, scarcely apiculate; antherodes very slender. Disk in ♂ flowers annular or patelliform, partly fused to base of staminal tube and ovary, in ♀ flowers a small swelling fused to the base of the ovary. Ovary 2–5(–6)-locular; locules with 2 superposed ovules; style short and stout; style-head capitate, with 2–5 indistinct stigmatic lobes; pistillode slender, sometimes with vestigial ovules. Drupe with 2–4(–6), 1(–2)-seeded pyrenes.

4 species in Africa.

Second-year branchlets usually less than 0.6 cm. in diameter, smooth, with scattered leaf-scars, closely lenticellate with large whitish lenticels; leaflets tapering to an acuminate or subacuminate apex; medium-sized tree of evergreen forest . . . .   1. *E. capensis*
Second-year branchlets usually more than 0.7 cm. in diameter, rough, with thick corky bark and crowded leaf-scars, lenticels inconspicuous; leaflets not tapering, apex broadly rounded to subtruncate, emarginate or with a minute apiculum; small tree of open woodland . . .   2. *E. benguelensis*

1. **E. capensis** *Sparrm.* in Sv. Vet. Akad. Handl. 40: 282, t. 9 (1779); C. DC. in A. & C. DC., Monogr. Phan. 1: 641 (1878); Chalk et al. in Chalk & Burtt Davy, For. Trees. Brit. Emp. 3: 51, t. 7/10 (1935); F. White & Styles in F.Z. 2: 316, t. 62 (1963); Styles in E. Afr. Agric. Journ. 39: 409 (1974); F. White in Bothalia 16: 155, fig. 5 (1986); F. White & Styles in F.S.A. 18(3): 51, fig. 16 (1986); Styles & F. White in Fl. Eth. 3: 489, fig. 124.6/6–9 (1990). Type: South Africa, *Sparrmann* (S, holo.!)

Medium-sized evergreen or semi-evergreen tree up to 30 m. tall, but usually less; bole slightly buttressed or fluted at base, up to 1 m. diameter at breast height; second-year branchlets slender, usually less than 0.6 cm. diameter, smooth, with scattered leaf-scars, closely lenticellate with large whitish lenticels. Leaves imparipinnate, usually entirely glabrous, sometimes pubescent, rarely tomentose; petiole and rhachis up to 35 cm. long; leaflets opposite or subopposite, 7–15, subsessile or shortly petiolulate, lanceolate to oblong-lanceolate, up to 14.5 × 6 cm., usually much smaller, tapering to an acuminate or subacuminate apex, base asymmetric, lower surface rarely drying whitish. Flowers white or pinkish white, sweet-scented, borne in many-flowered cymose panicles. Calyx 0.2 cm. long, sparsely to densely puberulous. Petals elliptic-oblong, 0.4–0.5 cm. long, densely puberulous on both surfaces. Staminal tube 0.2 cm. long, puberulous outside, densely bearded at throat inside. Ovary 0.15 × 0.2 cm., densely setulose; style 0.05–0.1 cm. long. Drupe ± 1.5 × 1.5 cm., deep red, with 2–4 pyrenes. Fig. 10/6–9, p 36.

UGANDA. Karamoja District: Lwala Forest Reserve, Feb. 1955, *Philip* 669!; Bunyoro District: Masindi, Feb. 1938, *Eggeling* 3504!; Mbale District: Siti, N. Elgon Forest Reserve, Jan. 1963, *Styles* 315! & 315A!
KENYA. Turkana District: Murua Nysigar Hills, Feb. 1965, *Newbould* 7127!; Embu District: Mt. Kenya, Irangi Forest Station, Aug. 1949, *White* 1094!; Teita Hills, Ngangao Forest, May 1972, *Faden et al.* 72/207!
TANZANIA. Mwanza District: Maisome I., Nov. 1954, *Carmichael* 459!; Lushoto District: W. Usambaras, Bumbuli, Nov. 1931, *Wigg* 25!; Rungwe District: Kyimbila, Dec. 1913, *Stolz* 2308!
DISTR. U 1–4; K 1–7; T 1–8; from Senegal to Ethiopia and southwards to Botswana and South Africa (Eastern Cape) but very local in Guineo-Congolian rain-forest
HAB. Montane, mid-altitude and riparian forest, often at edges; more rarely (Uganda) in woodland and wooded grassland; 600–2650 m.

SYN. *E. senegalensis* A. Juss., Mém. Mus. Nat. Hist. Nat., Paris 19: 234, 273 (1830); C.F.A. 1: 316 (1951); I.T.U., ed. 2: 174 (1952); Staner & Gilbert in F.C.B. 7: 208 (1958); F.W.T.A., ed. 2, 1: 705 (1958); Styles in E. Afr. Agric. Journ. 39: 410 (1974). Type: Senegal, *Leprieur* (P-JUSS, holo.!)
      Trichilia rueppelliana Fresen. in Mus. Senckenb. 2: 278 (1837). Type: Ethiopia, between Halei and Temben, *Rüppell* (FR, holo.!)
      Ekebergia rueppelliana (Fresen.) A. Rich., Tent. Fl. Abyss. 1: 105 (1847); F.T.A. 1: 333 (1868); C. DC. in A. & C. DC., Monogr. Phan. 1: 643 (1878); Gürke in P.O.A. C: 231 (1895); T.T.C.L.: 315 (1949); I.T.U., ed. 2: 174 (1952); Staner & Gilbert in F.C.B. 7: 208 (1958); K.T.S. 267, t. 54 (1961)
      E. petitiana A. Rich., Tent. Fl. Abyss. 1: 105 (1847). Type: Ethiopia, Tchelicot, *Quartin-Dillon & Petit* (P, holo.!)

E. meyeri C. DC. in A. & C. DC., Monogr. Phan. 1: 642 (1878); T.T.C.L.:315 (1949). Type: South
  Africa, *Drège* (B, holo.†)
E. buchananii Harms in E.J. 23: 164 (1896); T.T.C.L.: 314 (1949). Type: Malawi, *Buchanan* 39 (B,
  holo.†, K, iso.!)
E. complanata Bak.f. in J.L.S. 37: 132 (1905); I.T.U., ed. 2: 172 (1952). Type: Uganda, Mengo
  District, Buvuma I., *Bagshawe* 600 (BM, holo.!)
E. petitiana A. Rich. var. *australis* Bak.f. in J.L.S. 37: 133 (1905). Type: Uganda, Kigezi District,
  Rukiga, *Bagshawe* 466 (BM, holo.!)
E. holtzii Harms in E.J. 46: 160 (1911); T.T.C.L.: 315 (1949). Types: Tanzania, Dar es Salaam,
  Holtz 1023 (B, syn.†) & *Holtz* 1026 (B, syn.†)
E. mildbraedii Harms in N.B.G.B. 7: 229 (1917); Staner & Gilbert in F.C.B. 7: 210, fig. 23 (1958).
  Type: Cameroon, *Mildbraed* 8479 (B, holo.†, K, iso.!)
E. sp. nr. velutina Dunkley sensu Eggeling & Dale, I.T.U., ed. 2: 175 (1952)
E. sp. sensu Eggeling & Dale, I.T.U., ed. 2: 175 (1952)

NOTE.   *E. capensis* is somewhat variable, which is not surprising in view of its wide geographical
  distribution and considerable ecological versatility. Some variation is correlated with geography
  but only weakly so; there seems to be little justification for formal taxonomic division. In *E.
  senegalensis* the leaflets show less taper towards the apex but some specimens (e.g. *Onochie F.H.I.*
  8188, Nigeria) can be exactly matched in E. Africa. *E. mildbraedii* has more leaflets and more
  numerous lateral nerves than is usual for *E. capensis*, but there is appreciable overlap. The lower
  surface of the leaflets is usually glabrous but a tomentose variant (*E. buchananii*) occurs
  sporadically on the eastern side of Africa. The leaf-undersurface is sometimes whitish. This is due
  to prominent epidermal papillae. There is however continuous variation from smooth, through
  striate or weakly papillose, to strongly papillose cells (V.C. Uzoechina, pers. comm.); there are no
  pronounced geographical trends in this feature.
    The total range of variation in leaflet-shape and size of *E. capensis* in Southern Africa has been
  visually displayed by F. White in a published silhouette diagram (in Bothalia 16, fig. 5 (1986)).
  Nearly all specimens collected in East Africa can be matched in fig. 5, except that a few are slightly
  broader and more broadly rounded at the base. They cannot be confused with the leaflets of *E.
  benguelensis*, which unequivocally lie outside the limits of variation of *E. capensis*.
    Some atypical specimens of *E. capensis* (with broader, less tapering leaflets), e.g. *Greenway &
  Kanuri* 12024, have been misidentified as *E. benguelensis*. The bark on their branchlets, however, is
  smooth and lenticellate, and although the taper of the leaflets is less than in normal *E. capensis* it is
  much more pronounced than in *E. benguelensis*.

2. **E. benguelensis** C. DC. in A. & C. DC., Monogr. Phan. 1: 642 (1878); C.F.A. 1: 316
(1951); Staner & Gilbert in F.C.B. 7: 209 (1958); F. White & Styles in F.Z. 2: 318 (1963).
Type: Angola, Huila, Lopollo, *Welwitsch* 1301 (BM, holo.!)

Small semi-evergreen tree up to 10 m. tall, frequently stunted and of irregular growth;
bark rough, exfoliating in irregular scales; first-year branchlets often with smooth
reddish bark; second-year branchlets stout, usually more than 0.7 cm. in diameter, rough,
with thick corky bark and crowded leaf-scars; lenticels inconspicuous. Leaves
imparipinnate, glabrous to densely pubescent; petiole and rhachis up to 20 cm. long,
frequently reddish; leaflets opposite or subopposite, rarely alternate, 7–9, subsessile or
shortly petiolulate, ovate, ovate-oblong, oblong-elliptic or elliptic, up to 9 × 5 cm., usually
smaller, apex broadly rounded to subtruncate, emarginate or with a minute apiculum,
base usually asymmetric, lower surface nearly always drying whitish. Flowers white or
pinkish white, sweet-scented, borne in many-flowered cymose panicles. Calyx 0.15 cm.
long, sparsely puberulous to tomentellous. Petals elliptic-oblong, 0.5 × 0.25 cm., densely
puberulous on both surfaces. Staminal tube 0.2–0.3 cm. long, puberulous outside, densely
bearded at throat inside. Ovary 0.15 × 0.25 cm., densely setulose; style 0.05 cm. long. Drupe
± 1.5 × 1.5 cm., bright red, with 2–4 pyrenes.

TANZANIA.   Singida District: Mkalama, Oct. 1935, *B.D. Burtt* 5238!; Ulanga District: Selous Game
  Reserve, 6 km. N. of Mlahi, *Vollesen M.R.C.* 4739!; Songea District: Mpapa, Oct. 1956, *Semsei* 2527!
DISTR.   T 1, 4–8; Angola, southern Zaire, Zambia, Malawi, Mozambique, Zimbabwe
HAB.   Woodland and wooded grassland; 275–1900 m.

SYN.   *E. arborea* Bak.f. in J.B. 37: 427 (1899); T.T.C.L.: 314 (1949). Type: Zimbabwe, Harare [Salisbury],
    Rand 612 (BM, holo.!)
    E. sclerophylla Harms in E.J. 28: 415 (1900); T.T.C.L.: 315 (1949). Type: Tanzania, Iringa District,
    Uhehe, Ifunda [Funda], *Goetze* 721 (B, holo.†, BR, lecto.! here designated)
    E. nana Harms in N.B.G.B. 11: 401 (1932); T.T.C.L.: 315 (1949). Types: Tanzania, Njombe
    District, Lupembe, Likanga, *Schlieben* 470 (B, syn.†, BR, isosyn.!) & *Schlieben* 471 (B, syn.†)
    E. sp. nr. benguelensis C. DC. sensu T.T.C.L.: 315 (1949)

NOTE.   Relationships with *E. capensis* are discussed under that species.

8. TURRAEANTHUS

Baillon in Adansonia 11: 261 (1874) & Hist. Pl. 5: 500 (1874–75); Pennington & Styles in Blumea 22: 493 (1975)

Trees or treelets. Indumentum of simple hairs. Leaves pari- or imparipinnate. Flowers unisexual (plants dioecious). Calyx almost entire or shallowly or irregularly lobed. Petals 4–5(–6), valvate, fused (to the middle or beyond) to the staminal tube. Staminal tube cylindrical, slightly expanded at the mouth, margin crenate or shallowly lobed; anthers 8–12, in a single irregular whorl inserted within the throat of the staminal tube, completely included or partly exserted. Disk absent. Ovary 4–5-locular; locules with 2 superposed or oblique ovules; style-head discoid, with a slight central depression. Fruit a leathery, 3–5-valved, loculicidal capsule. Seeds with a fleshy aril.

3 species in tropical Africa.

T. africanus (*C. DC.*) *Pellegrin* in Not. Syst. 2: 16, 68 (1911); C.F.A. 1: 317 (1951); F.W.T.A., ed. 2, 1: 707 (1958); Staner & Gilbert in F.C.B. 7: 200 (1958); Styles in E. Afr. Agric. Journ. 39: 422 (1974). Type: Angola, Golungo Alto, Bango, mata de Quisucula, *Welwitsch* 1306 (LISU, holo., BM, K, iso.!)

A poorly shaped, evergreen, understorey tree up to 21 m. tall (elsewhere up to 35 m. or more) with spreading crown and dense foliage; bole irregular, branched low down; bark grey or whitish, shallowly fissured, scaling in small pieces. Leaves up to 60 cm. long; petiole flattened or grooved on upper surface, especially towards the base; leaflets oblong-elliptic or oblanceolate-oblong, 20–36, up to 22 × 4 cm., apex usually with a short acumen, base asymmetric; lateral nerves in 15–25 pairs, departing from the midrib at a very wide angle; lower surface soon becoming glabrous. Flowers club-shaped in bud, in panicles up to 30 cm. long in leaf-axils and on older branchlets. Calyx cupular, 5-toothed. Petals up to 2.5 cm. long, rusty-tomentose outside, free from each other but united for about half of their length with the staminal tube. Fruit yellow or orange, shaped like a fig, up to 3.5 cm. long. Seeds with a whitish aril. Fig. 11.

UGANDA. Toro District: Itwara Forest Reserve, Feb. 1943, *St. Clair-Thompson* in *Eggeling* 5229! & Apr. 1943, *Eggeling* 5274! & Kibale Forest Reserve, Nov. 1963, *Styles* 244!
DISTR. U 2; Guineo-Congolian rain-forest from Sierra Leone to Uganda and Angola
HAB. In Uganda only known from the understorey of Itwara and Kibale forests where it occurs fairly frequently in riparian and poorly drained areas; 1525 m.
SYN. *Guarea africana* C. DC. in A. & C. DC., Monogr. Phan. 1: 576 (1878)
Turraeanthus sp. sensu Eggeling & Dale, I.T.U., ed. 2: 201 (1952)
NOTE. In West Africa the timber is highly regarded for cabinet-making but in Uganda the trees are too small and too poor in shape to be of any commercial importance.

9. GUAREA

L., Mant. Pl. Alt.: 150, 228 (1771); Pennington & Styles in Blumea 22: 494 (1975), *nom. conserv.*

Trees or treelets. Indumentum of simple hairs. Leaves pinnate, with a dormant terminal bud or more rarely (sometimes in E. Africa) a terminal leaflet. Flowers unisexual (plants dioecious). Calyx (in E. Africa) irregularly lobed. Petals 5 and free from the staminal tube (in E. Africa). Staminal tube cylindric, ending in very short, slightly emarginate appendages; anthers 10 (in E. Africa), inserted in the throat of the staminal tube and completely included. Disk annular at the base of the ovary. Ovary 3–5-locular; locules with 1 or 2 ovules; style-head discoid with a small central depression. Fruit (in E. Africa) either a leathery capsule with 3–5 one-seeded valves, or a large leathery cleistocarp. Seeds completely covered by a fleshy orange or yellow aril.

5 species in tropical Africa and 35 in the Neotropics.

Petiole up to 4 cm. long; petals up to 0.7 cm. long; fruit a capsule
up to 5.5 cm. in diameter . . . . 1. *G. cedrata*
Petiole usually 4.5–8.5 cm. long; petals 1.3–1.7 cm. long; fruit
indehiscent, (8–)10–12 cm. in diameter . . . . 2. *G. mayombensis*

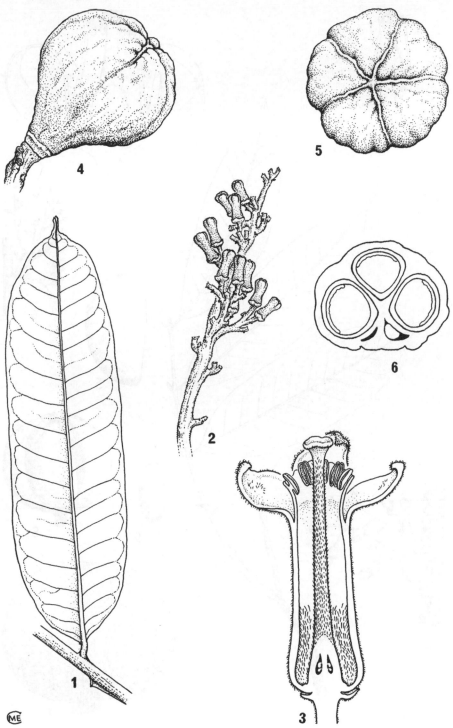

FIG. 11. *TURRAEANTHUS AFRICANUS*—1, leaflet, × ⅖; 2, part of inflorescence, × ⅖; 3, half of ♂ flower, vertical section, × 4; 4,5, two views of fruit, × 1; 6, fruit, transverse section, × 1. 1, from *Styles* 244; 2, from *Brenan* 9348; 3, from *Jacques-Félix* 2539; 4–6, from *Hawthorne* s.n. Drawn by Maureen Church.

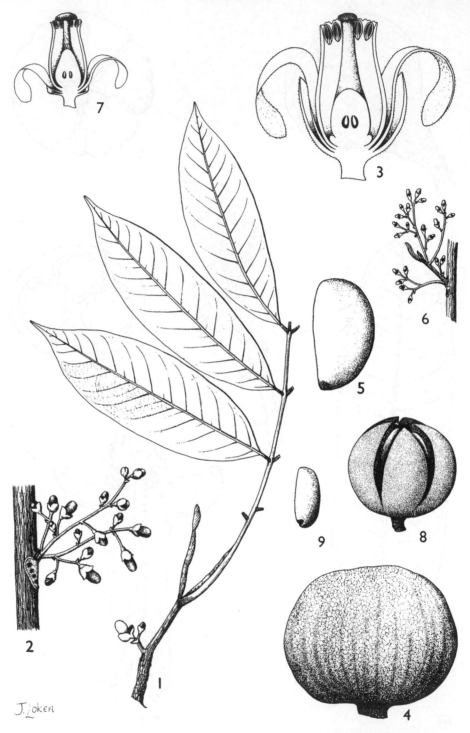

FIG. 12. *GUAREA MAYOMBENSIS*—1, flowering branchlet, × ⅓; 2, inflorescence, × ⅓; 3, half of ♂ flower, vertical section, × 2⅔; 4, cleistocarp, × ½; 5, seed, × ⅓. *G. CEDRATA*—6, inflorescences, × ½; 7, half of ♀ flower, vertical section, × 2⅔; 8, capsule, × ½; 9, seed, × ⅓. 1–3, from *Bafendaye* K1/34; 4,5, from *Styles* s.n.; 6, from *Toussaint* 2433; 7, from *Kennedy* 309; 8,9, from *Louis* 2868. Drawn by Julia Loken

1. **G. cedrata** (*A. Chev.*) *Pellegrin* in Bull. Soc. Bot. Fr. 75: 480 (1928); I.T.U., ed. 2: 184 (1952); F.W.T.A., ed. 2, 1: 706 (1958); Staner & Gilbert in F.C.B. 7: 204 (1958); Styles in E. Afr. Agric. Journ. 39: 412 (1974), excl. *Styles* 102. Lectotype designated here: Ivory Coast, Bouroukrou, *Chevalier* 16125 (P, lecto.!)

Tree up to 45 m. tall, with a dense rounded crown; bole slender, fluted above, with short blunt buttresses below; bark smooth, grey, sometimes with a reddish tinge, often with concentric rings of lenticels forming 'mussel shell markings' where scales have fallen. Leaves paripinnate or sometimes imparipinnate, up to 20 cm. long, pinkish when young, in herbarium specimens usually drying brown; petiole up to 4 cm. long, deeply grooved and with broadly winged margins; leaflets 8–12, often galled, narrowly oblong-elliptic, up to 20 × 5(–8) cm., apex acuminate, base cuneate, rarely rounded, often asymmetric; margin often undulate; lateral nerves usually in 10–12 pairs, very prominent beneath; lower surface glabrous or almost so, venation closely reticulate. Flowers pale yellow, fragrant, in panicles up to 7 cm. long in the axils of fallen leaves or on young shoots below the developing leaves. Calyx 0.1–0.2 cm. long, tomentellous. Petals up to 0.7 cm. long, tomentellous. Capsule subglobose, up to 5.5 cm. in diameter, pinkish velutinous, with 3–5 leathery valves. Seeds 3–5, up to 4 × 2 × 2 cm., completely covered by a fleshy orange aril. Fig. 12/6–9.

UGANDA. Bunyoro District: Budongo Forest, Sept. 1962, *Styles* 104!; Toro District: Bwamba Forest, Aug. 1937, *Eggeling* 3373!; Mengo District: Lake Victoria, Damba I., May 1948, *Sangster* 1018!
DISTR. U 2, 4; Sierra Leone to Gabon and eastwards to Uganda
HAB. Lowland rain-forest; ± 1100 m.

SYN. *Trichilia cedrata* A. Chev. in Vég. Ut. Afr. Trop. Fr. 5: 214 (1909)

NOTE. The tree produces a high-grade scented timber suitable for furniture and cabinet making. It resembles African mahogany (*Khaya*) in colour and texture but is too rare and too thinly distributed in Uganda to be of any great commercial use.
  G. cedrata is easily confused with *Blighia welwitschii* (Hiern) Radlk. (Sapindaceae; syn. *B. wildemaniana* De Wild.). Eggeling & Dale (I.T.U., ed. 2: 377) say that their leaves are indistinguishable. *Blighia* differs however in having much longer petioles which are flattened on the upper surface and only narrowly flanged.

2. **G. mayombensis** *Pellegrin* in Bull. Mus. Hist. Nat., Paris 27: 449 (1921), Fl. Mayombe 1: 54 (1924) & in Bull. Soc. Bot. Fr. 86: 152 (1939); I.T.U., ed. 2: 184 (1952). Type: Gabon, Tchibanga, *Le Testu* 1990 (P, holo.!, K, iso.!)

Tree up to 25 m. tall; bark grey, rough, Leaves pari- or imparipinnate, up to 60 cm. long, but usually smaller; petiole deeply grooved and broadly winged, usually 4.5–8.5 cm. long; leaflets up to 15, lanceolate to oblanceolate, up to 44 × 10 cm. but usually much smaller, tapering more gradually to the acuminate or subacuminate apex than in *G. cedrata*, and with the lateral nerves and venation much less prominent. Flowers similar to those of *G. cedrata*, but much larger and in fewer-flowered inflorescences. Cleistocarp, up to 12 cm. in diameter, indehiscent or breaking up after falling. Seeds up to 9 × 5 cm., covered by a thin yellow aril. Fig. 12/1–5.

UGANDA. Kigezi District: Ishasha Gorge, Feb. 1945, *Greenway & Eggeling* 7097! & Kayonza Forest, Aug. 1971, *Hamilton* 71/164! & Kayonza Forest, Feb. 1963, *Zebukoza* in *Styles* 352!
DISTR. U 2; Gabon, W. and E. Zaire (but apparently absent from the Zaire basin)
HAB. Mid-altitude rain-forest; 1220–1600 m.

SYN. *Leplaea mayombensis* (Pellegrin) Staner in B.J.B.B. 16: 204 (1941); Staner & Gilbert in F.C.B. 7: 212 (1958); Styles in E. Afr. Agric. Journ. 39: 415 (1974)

10. **CEDRELA**

P. Browne, Hist. Jamaica: 158, t. 10/1 (1756); L., Syst. Nat., ed. 10: 940 (1759); Pennington & Styles in Blumea 22: 512 (1975); Styles in Pennington, Fl. Neotrop. Monogr. 28: 360 (1981)

Deciduous trees. Leaves usually paripinnate; leaflets entire. Flowers 5-merous, unisexual (plants monoecious) in large much-branched panicles. Calyx variable, lobed almost to the base, shallowly dentate, or cup-shaped and split down one side. Petals 5,

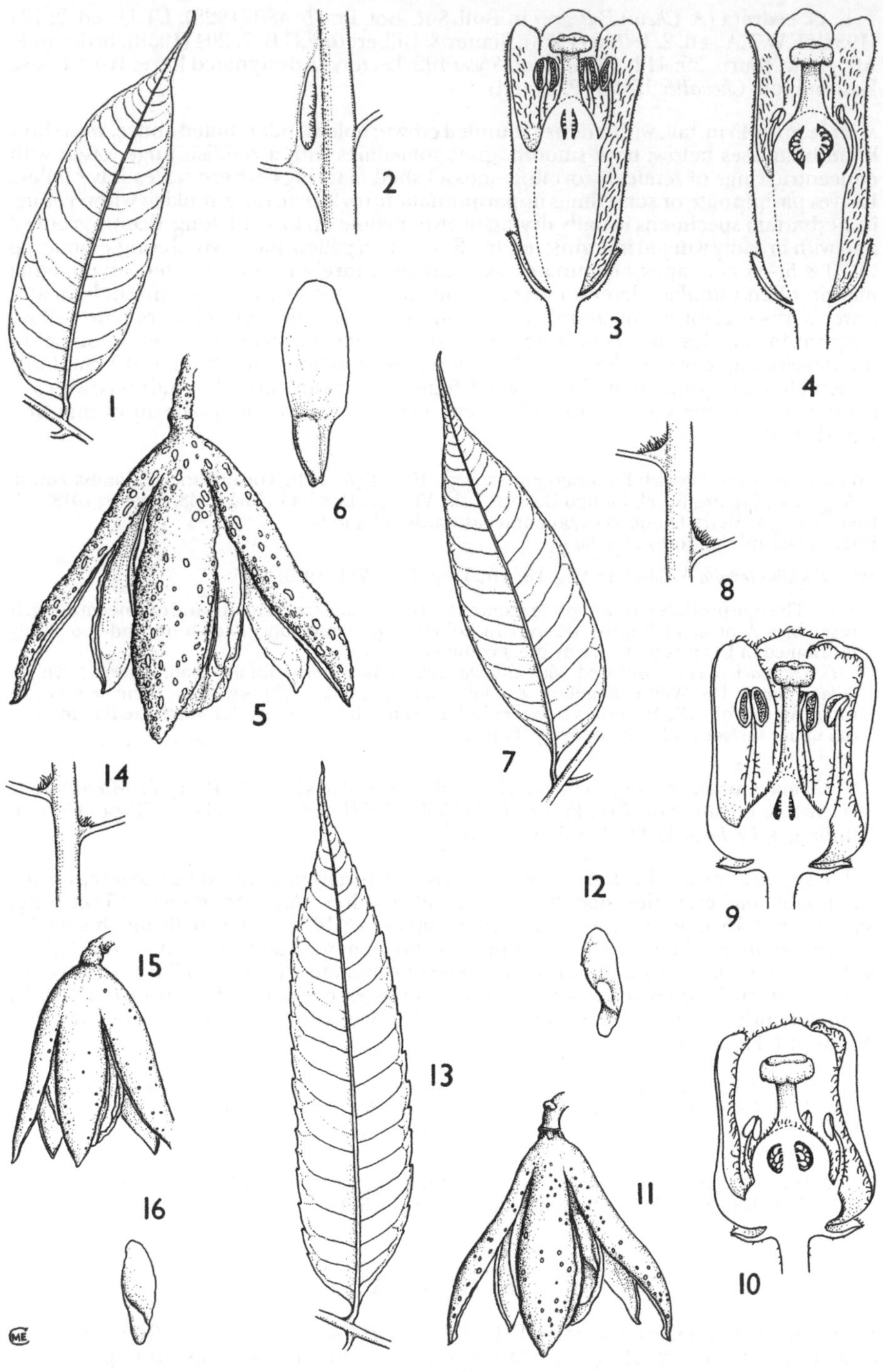

FIG. 13. *CEDRELA ODORATA*—**1**, leaflet, × ½; **2**, domatium, × 5¾; **3**, half ♂ flower, vertical section; × 4⅘; **4**, half ♀ flower, vertical section, × 4⅘; **5**, capsule, × 1; **6**, seed, × 1. *TOONA CILIATA*—**7**, leaflet, × ½; **8**, domatia, × 5¾; **9**, half ♂ flower, vertical section, × 4⅘; **10**, half ♀ flower, vertical section, × 4⅘; **11**, capsule, × 1⅔; **12**, seed, × 1. *T. SERRATA*—**13**, leaflet, × ½; **14**, domatia, × 5¾; **15**, capsule, × 1⅔; **16**, seed, × 1. 1,2, from *Styles* 19; 3,4, from *Pennington* 9650; 5,6, from *Styles* 166; 7–12, from *Styles* 266; 13,14, from *Styles* 289; 15,16, from *Dale* U439. Drawn by Maureen Church.

imbricate, adnate for one-third of their length to the long columnar androgynophore by a median carina (thereby preventing their spreading in open flowers). Stamens 5, free, adnate to the androgynophore below; staminodes absent. Ovary 5-locular; locules with 6–12 ovules; style-head discoid with glandular stigmatic papillae. Fruit a pendulous, thinly or thickly woody, obovoid or calviform, septifragal capsule, opening from the apex by 5 valves; columella woody, sharply 5-angled, extending to the apex of the capsule; seed-scars conspicuous. Seeds with a terminal wing, attached by the seed-end to the apex of the columella.

About 8 species in the Neotropics. *C. odorata* is widely planted in Africa for timber.

16. **C. odorata** *L.*, Syst. Nat., ed. 10: 940 (1759); Dale, Introd. Trees Uganda: 21 (1953); Styles in Pennington & Styles, Fl. Neotrop. Monogr. 28: 374, fig. 76, 76a (1981); F. White & Styles in F.S.A. 18(3): 60 (1986). Type: plate 10/1 of P. Browne, Hist. Jamaica: 158 (1756)

Tree up to 30 m. tall. Leaflets usually 12 to 28, entire, oblong-lanceolate, up to 15 × 6.5 cm., rather suddenly tapered to the relatively short and broad acumen; proximal lateral nerves on lower surface with conspicuous elongate axillary pockets (domatia) with minute marginal hairs or glabrous. Inflorescence up to 50 cm. long. Capsule robust, usually 3–4.2 cm. long; valves with conspicuous lenticels. Fig. 13/1–6.

UGANDA. Mengo District: Kampala, Makerere College, July 1964, *D.A.H. Taylor*!
KENYA. C. Kavirondo District: Maseno, *District Forest Officer* in *E.A.H.* 176/54!
TANZANIA. Ngara township, cult. Nov. 1957, *Willan* 310!; Moshi township, cult. Dec. 1951, *McCoy-Hill* 21!; Kigoma District: Ujiji, *Wigg* 452!
DISTR. Extensively grown in trial plots in East Africa because of its potential importance as a timber tree; widespread in the Neotropics from Mexico and the West Indies to N. Argentina

SYN. *C. mexicana* M. Roemer, Synops. Monogr. Hesp. 1: 137 (1846); T.T.C.L.: 314 (1949). Type: Mexico, Papantla, *Deppe & Schiede* 1304 (LE, holo., K, NY, iso.!)
C. sp. sensu Brenan, T.T.C.L: 314 (1949)

NOTE. In its native lands, *Cedrela odorata* produces the famous Spanish Cedar of commerce, a timber which is highly regarded for joinery of all kinds including cabinet making. It was formerly much favoured for making cigar boxes.

11. **TOONA**

M. Roemer, Synops. Monogr. Hesp. 1: 131, 139 (1846); Pennington & Styles in Blumea 22: 512 (1975)

Cedrela P. Browne sect. *Toona* Endl., Gen. Pl. 2: 1055 (1840)

Deciduous or semi-evergreen trees. Leaves paripinnate; leaflets entire or toothed. Flowers 5-merous, unisexual (plants monoecious), in large much-branched panicles. Sepals free or partly united. Petals 5, imbricate, adnate at the base by a median carina to the short, cushion-shaped androgynophore. Stamens 5, inserted on the androgynophore; 1–5 thread-like staminodes sometimes present. Ovary 5-locular; locules with 6–10 ovules; style-head discoid with glandular stigmatic papillae. Fruit a pendulous, thinly woody, ellipsoid or obovoid, septifragal capsule, opening from the apex by 5 valves, columella softly woody, 5-angled, extending to the apex of the capsule. Seeds either winged at both ends and attached towards the apex of the columella, or with a single wing and then attached by the seed-end to the base of the columella.

Two or more species in the Old World eastwards from India to China and Australia.

Margin of leaflets entire; disk and ovary puberulous; seeds
    winged at both ends . . . . 1. *T. ciliata*
Margin of leaflets serrate; disk and ovary glabrous; seeds
    winged at one end only . . . . 2. *T. serrata*

1. **T. ciliata** *M. Roemer*, Synops. Monogr. Hesp. 1: 139 (1846); T.T.C.L: 318(1949); Dale, Introd. Trees Uganda: 69 (1953); Styles in E. Afr. Agric. Journ. 39: 417 (1974); F. White & Styles in F.S.A. 18(3): 61 (1986). Type: India, Madras, *Roxburgh* in *Herb. Willd.* 4828 (B-WILLD, holo., fide D.J. Mabberley, pers. comm.)

Tree up to 25 m. tall. Leaflets 10–24, entire, lanceolate or lanceolate-elliptic, up to 15×5 cm., gradually tapering from near the base to the long, narrow, attenuate-acuminate apex; proximal lateral nerves on lower surface with minute deltate axillary pockets (domatia) with minute apical hairs. Inflorescence up to 35 cm. long. Capsule delicate, 1.5–2 cm. long, valves without conspicuous lenticels. Fig. 13/7–12.

UGANDA. Acholi District: Opaka Local Forest Reserve, cult. Dec. 1962, *Styles* 266!; Mengo District: Entebbe Botanic Garden, cult. Sept. 1962, *Styles* 17!
KENYA. Nairobi, Davidson's Garden, cult. Oct. 1964, *Gillett* 16284!
TANZANIA. Tanga District: Amani–Sigi, cult. Sept. 1958, *Willan* 348!; Buha District: Marera Mission, cult. Apr. 1964, *Pirozynski* 642!
DISTR. Extensively planted in East Africa (and other parts of Africa) as an avenue tree, for ornament, fuel and experimental forestry purposes; in some countries, e.g. Malawi, it has become extensively naturalized; native of Asia and Australia.

SYN. *Cedrela toona* Rottler & Willd. in Ges. Nat. Fr. Berlin, Neue Schr. 2: 198 (1803). Type as for *Toona ciliata*
   C. australis F. Muell., Fragm. 1: 4 (1858). Type: E. Australian coast, no specimen cited
   Toona australis (F. Muell.) Harms in E. & P. Pf. III. 4: 270 (1896)
   Cedrela toona Rottler & Willd. var. *australis* C.DC. in Rec. Bot. Surv. India 3: 368 (1908). Type as for *C. australis* F. Muell.

NOTE. This species has been frequently confused with the superficially similar *Cedrela odorata*. They are however readily distinguished by the differences given above in the generic and specific diagnoses.
   The name *Toona ciliata* has been widely used, both in botanical works and in the general, especially dendrological, literature; it is well-known and in current use throughout the tropics. Its homotypic synonym, *Cedrela toona*, features prominently in the older literature. There is at least one earlier but comparatively little-known epithet belonging to a species which may be considered conspecific with *Toona ciliata* by some workers. In order to prevent the nomenclatural instability which would then ensue, we recommend that *Toona ciliata* should be conserved.

2. **T. serrata** (*Royle*) *M. Roemer*, Synops. Monogr. Hesp. 1: 139 (1846); Dale, Introd. Trees Uganda: 69 (1953); Styles in E. Afr. Agric. Journ. 39: 417 (1974). Type: India, *Royle* (K, holo.!)

Tree up to 30 m. tall; bark rough, brownish black, deeply fissured. Leaflets 18–40, serrate, narrowly oblong or oblong-lanceolate, up to 20 × 4 cm., apex acuminate, base obliquely rounded, very asymmetric; midrib often reddish; proximal lateral nerves on lower surface with minute deltate axillary pockets (domatia) with minute apical hairs. Inflorescence pendulous, up to 90 cm. long. Capsule larger and coarser than in *T. ciliata* ± 2.5 cm. long; valves smooth or sometimes with small, inconspicuous lenticels. Fig. 13/13–16, p. 44.

UGANDA. Busoga District: Jinja Cemetery, cult. Sept. 1945, *Dale* U439! & Jinja Market, cult. Sept. 1945, *Dale* U440!
TANZANIA. Lushoto District: E. Usambara Mts., Amani, cult. Jan. 1931, *Greenway* 2855! & Apr. 1958, *Parry* 247!
DISTR. Cultivated in Uganda and Tanzania as an avenue tree and in windbreaks and plantations; native of the Himalayas

SYN. *Cedrela serrata* Royle, Ill. Bot. Himal.: 144, t. 25C (1834-5); T.T.C.L.: 314 (1949)

NOTE. According to Mabberley (Tree Flora of Malaya 4: 256 (1989)) *T. serrata* (Royle) M. Roemer is a synonym of *T. sinensis* (A. Juss.) M. Roemer.

12. KHAYA

A. Juss. in Mém. Mus. Nat. Hist. Nat., Paris 19: 249 (1830); Pennington & Styles in Blumea 22: 515 (1975)

Large trees. Leaves paripinnate. Flowers 4- or 5-merous, unisexual (plants monoecious) in large much-branched panicles. Calyx lobed almost to the base, the lobes suborbicular, imbricate. Petals 4 or 5, free, contorted, erect in open flowers, somewhat hooded. Staminal tube urceolate, bearing 8–10 included anthers or antherodes towards the apex and terminated by 8–10 suborbicular, emarginate or irregularly lobed, overlapping appendages alternating with the anthers or antherodes. Disk in ♂ flowers cushion-shaped, fused to the base of the pistillode but free from the base of the staminal tube, less conspicuous in ♀ flowers. Ovary 4–5-locular, each locule with 12–18 ovules; style-head discoid with crenulate margin, upper surface with 4–5 radiating stigmatic ridges. Fruit an erect, subglobose, woody septifragal capsule opening by 4–5(–6) valves from the apex, the valves remaining joined at the base and often with rough, fibrous marginal strands; columella not extending to the apex of the capsule, with 4–5(–6) sharp, hard, woody ridges; seed-scars white, conspicuous. Seeds 8–18 per locule, broadly transversely ellipsoid or suborbicular, narrowly winged all round the margin.

6 species in tropical Africa, Madagascar and the Comores.

The species of *Khaya* are very uniform in floral and fruit morphology, and (other than in habit, distribution and ecology) differ principally in details of leaflet outline and minor features of the fruit. The East African species show almost complete ecogeographical replacement.

Greenway (in E. Afr. Agric. Journ: 8–14 (1947)) has published an interesting account of the East African Khayas, dealing especially with the history of their exploitation and afforestation. When he wrote, 9 of the 16 previously published names in *Khaya* were in use on the African mainland, 4 of them in East Africa. Greenway suggested that further information might lead to further reduction. In our opinion *Khaya nyasica* cannot be kept apart from *K. anthotheca*, and the status of *K. grandifoliola* is doubtful. *K. anthotheca* is the most important indigenous timber in Uganda and there is a copious literature dealing with it and the other two Uganda species. All three reach their geographical limits in Uganda, where their taxonomy, especially the relationships of *K. anthotheca* and *K. grandifoliola*, is far from clear. We have attempted to follow traditional usage, but our treatment should be regarded as provisional. Putative hybrids between *K. anthotheca* and *K. grandifoliola* have been mentioned in the literature (e.g. by Eggeling & Harris (For Trees Brit. Emp. 4 (1939)), but different accounts give different interpretations. Looking at Africa as a whole, it seems that *K. anthotheca* and *K. grandifoliola* may not be specifically distinct, though some further checking is still required.

Leaflets mostly ovate-elliptic to elliptic, venation on lower surface (when dry) darker than the lamina, usually sunken; capsule 4–5-valved, often uniformly dark chocolate-brown inside, valves ± 0.45 cm. thick   .   .   .   1. *K. anthotheca*

Leaflets mostly oblong or oblong-elliptic, tapering much more abruptly to the apex and base than in *K. anthotheca*; venation on lower surface usually paler than or the same colour as the lamina, not sunken; capsule pale or reddish brown inside, with a pattern of darker marks:

Leaflets usually less than 8 × 5 cm., apex rounded or with a very short apiculum; capsule normally 4-valved, valves ± 4.5 cm. long and 0.3 cm. thick   .   .   .   .   .   .   2. *K. senegalensis*

Leaflets usually more than 12 × 5 cm., with a conspicuous, often twisted, apiculum; capsule normally 5-valved, valves ± 6 cm. long and 0.8 cm. thick   .   .   .   .   .   3. *K. grandifoliola*

1. **K. anthotheca** (*Welw.*) *C.DC.* in A. & C. DC., Monogr. Phan. 1: 721 (1878); V.E. 3(1): 803 (1915); Eggeling & Harris in For. Trees Brit. Emp. 4: 57, fig. 9, pl. ix (1939); T.T.C.L.:316 (1949); C.F.A. 1: 307, t. 13B (1951); I.T.U., ed. 2: 185, fig. 41 (1952); Staner & Gilbert in F.C.B. 7: 176 (1958); F.W.T.A., ed. 2, 1: 699 (1958); Styles in E. Afr. Agric. Journ. 39: 413 (1974). Type: Angola, Golungo Alto, Mont de Queta, *Welwitsch* 1314 (LISU, holo., BM, iso.!)

Large tree up to 60 m. tall, sometimes with very prominent surface roots; bole of large trees markedly buttressed to a height of 6 m., straight or with a slight 'wave', reaching a considerable height before branching (inside tall forest) or branched low down (in riparian forest), up to 4 m. in diameter above the buttresses; bark smooth but exfoliating in small circular scales ± 3 cm. in diameter and leaving a characteristically pock-marked, mottled grey and brown surface. Leaves up to 30(–40) cm. long, glabrous; leaflets 6–10(–14), coriaceous, mostly ovate-elliptic, or elliptic, up to 17 × 7 cm., gradually tapering to the cuspidate apex, base cuneate or obtuse, slightly asymmetric; petiolules 0.5–1.5 cm. long. Inflorescence an axillary panicle up to 25(–40) cm. long. Flowers sweet-scented, white, with a reddish disk surrounding the base of the ovary. Calyx 0.1–0.15 cm. long, lobed almost to the base; lobes suborbicular, ciliate. Petals elliptic, somewhat hooded, up to 0.6 × 0.3 cm. Staminal lobe ± 0.5 cm. long. Ovary 0.2 cm. in diameter; style less than 0.1 cm. long. Capsule 4–6 cm. in diameter, dehiscing by 4–5 valves. Seeds 1–2 × 1.5–3 cm., 6–12 per locule. Fig. 14, p. 48.

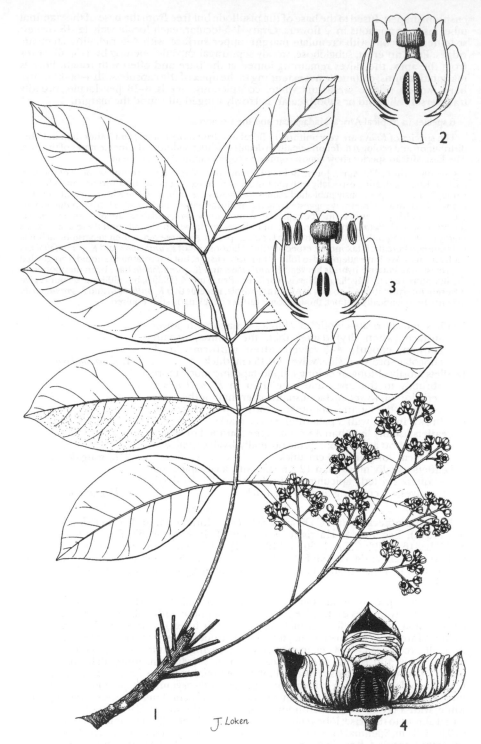

FIG. 14. *KHAYA ANTHOTHECA*—**1**, flowering shoot, × ⁹/₂₀; **2**, half ♀ flower, vertical section, × 4⁷/₁₀; **3**, half ♂ flower, vertical section, × 4⁷/₁₀; **4**, dehisced capsule, one valve removed, × ⁹/₂₀. 1, from *Styles* 339; 2,3, from *Styles* 399; 4, from *Styles* 337. Drawn by Julia Loken.

UGANDA. Bunyoro District: Budongo Forest, Oct. 1962, *Styles* 119!, Jan. 1963, *Styles* 338! & 339!
TANZANIA.   Lushoto District: E. Usambara Mts., Sigi, Oct. 1943, *Greenway* 6819! Morogoro District: Kimboza Forest Reserve, July 1952, *Semsei* 758!; Mbeya District: Mbozi, Kantasia Farm, Oct. 1936, *B.D. Burtt* 6243!
DISTR.   U 2; T 3–8; also widely grown in plantations in Tanzania and formerly used in enrichment planting in S. Mengo forests, Uganda; from Sierra Leone eastwards to Uganda and Tanzania and southwards to Angola, Zambia, Malawi, Mozambique and Zimbabwe but absent from the wettest forests of Upper and Lower Guinea, where it is replaced by *K. ivorensis* A. Chev.
HAB.   Lowland rain-forest and riparian forest; 120–1525 m.
SYN.   *Garretia anthotheca* Welw. in Ann. Conselho Ultram. 1858: 587 (1859), as '*anthoteca*'.
   Khaya nyasica Bak. f. in J.L.S. 40: 42, t. 1/4 (1911); V.E. 3(1): 803 (1915); T.T.C.L.: 317 (1949); Staner & Gilbert in F.C.B. 7: 178, fig. 21 (1958); F. White & Styles in F.Z. 2: 287, t. 54 (1963). Type: Mozambique, Lower Umswirizwi R. and Chirinda Forest, *Swynnerton* 15 (BM, holo.!, K, SRGH, iso.!)
NOTE.   One of the most valuable trees in Uganda where it formerly provided almost half the converted timber from Budongo Forest. It is used in high-class cabinet work and for the production of veneers.

2. **K. senegalensis** (*Desr.*) *A. Juss.* in Mém. Mus. Nat. Hist. Nat., Paris 19: 250, t. 21 (1830); V.E. 3(1): 801, fig. 376 (1915); Eggeling & Harris in For. Trees Brit. Emp. 4: 68, fig. 11, pl. xi (1939); I.T.U., ed. 2: 189, fig. 43 (1952); F.W.T.A., ed. 2, 1: 698 (1958); Styles in E. Afr. Agric. Journ. 39: 413 (1974). Type: Senegal, *Roussilon* (P-LAM, holo.!)

Tree up to 20 m. tall; bole often crooked; buttresses short or absent; crown rounded. Apart from the characters given in the key, very similar to *K. anthotheca* and *K. grandifoliola*, but usually smaller in all its parts. Leaves up to 25 cm. long, bright green and shining above, pale greyish green beneath; leaflets 4–10. Panicles up to 20 cm. long. Capsule up to 5 cm. in diameter.

UGANDA.   W. Nile District: Ofude, Feb. 1934, *Eggeling* 1499! & 16 km. Moyo to Laropi, Dec. 1962, *Styles* 270!; Acholi District: 32 km. Gulu to Kitgum, Dec. 1962, *Styles* 256!
DISTR.   U 1; Senegal eastwards to northern Uganda
HAB.   In savanna woodland, often in rocky places, and riparian, sometimes with *K. grandifoliola;* Eggeling & Harris, loc. cit.: 70; 915–1525 m.
SYN.   *Swietenia senegalensis* Desr. in Lam., Encycl. Méth. Bot. 3: 679 (1791)
NOTE.   The timber is similar to that of the other species though much heavier. It has been used for temporary buildings and rough furniture.

3. **K. grandifoliola** *C. DC.* in Bull. Soc. Bot. Fr. 54, Mém. 8: 10 (1907); V.E. 3(1): 803 (1915); Eggeling & Harris, For. Trees Brit. Emp. 4: 64, fig. 10, pl. x (1939); I.T.U., ed.2: 187, fig. 42 (1952); Staner & Gilbert in F.C.B. 7: 177 (1958); F.W.T.A., ed. 2, 1: 699, fig. 194 (1958); Styles in E. Afr. Agric. Journ. 39: 413 (1974). Type: Central African Republic, Tomi R., *Chevalier* 5769 (P, holo.!)

Tree usually less than 20(but occasionally up to 30) m. tall; bole unbuttressed or only slightly buttressed. Apart from the characters given in the key, very similar to *K. anthotheca* and *K. senegalensis.* Leaves up to 50 cm. long; leaflets 6–10, chartaceous. Panicles up to 35 cm. long. Capsule up to 7 cm. in diameter.

UGANDA.   W. Nile District: Koboko, Feb. 1934, *Eggeling* 1500! & Otzi Forest Reserve, 8 km. Moyo to Laropi, Dec. 1962, *Styles* 282!; Acholi District: 8 km. Kitgum to Gulu, Dec. 1962, *Styles* 267!
DISTR.   U 1; Guinea Bissau eastwards to northern Uganda
HAB.   In riparian forest; 765–1220 m.
NOTE.   The timber is considered suitable for cabinet-making and decorative joinery, but because the tree tends to branch low down the amount available is rather small.

13. ENTANDROPHRAGMA

C. DC. in Bull. Herb. Boiss. 2: 582, t. 21 (1894); Pennington & Styles in Blumea 22: 518 (1975)

Large trees. Leaves basically paripinnate, but leaflets sometimes alternate and leaves appearing imparipinnate. Flowers 5-merous, unisexual (plants dioecious) in large panicles. Calyx ± entire and cupuliform or with 5 acute lobes, aestivation open. Petals 5, free, contorted. Staminal tube urceolate or cup-shaped, the margin entire or shallowly or deeply lobed, with 10 shortly stalked anthers or antherodes on the margin of the tube or its lobes; appendages absent. Disk cushion-shaped, fused to the base of the ovary or pistillode, but free from the staminal tube and connected to it by 10 or 20 short ridges or partitions. Ovary 5-locular, each locule with 4–12 ovules; style-head discoid, with 5 radiating stigmatic lobes. Fruit a pendulous, elongate, woody, cigar-shaped, cylindrical or club-shaped, septifragal capsule, opening by 5 valves from the apex or base or from both simultaneously; columella softly woody, extending to the apex of the capsule, 5-angled or 5-ridged, deeply indented with the imprints of the seeds; seed-scars conspicuous or inconspicuous. Seeds with a terminal wing, 3–9 per locule, attached by the seed-end to the distal part of the columella and winged towards the base of the capsule.

11 species in tropical Africa.

Valves of capsule dehiscing from the apex and base and falling
    singly from columella . . . .   1. *E. cylindricum*
Valves of capsule dehiscing from the apex or base:
    Valves of capsule dehiscing from apex, persistent at base:
        Capsule with large rusty lenticels outside, valves shining
            golden-brown inside, with darker streaks   . . .   2. *E. utile*
        Capsule with small buff lenticels outside, valves pale buff
            inside with indistinct markings . . . .   3. *E. bussei*
    Valves of capsule dehiscing from the base and falling from
        the apex in the form of a calyptra:
        Leaflets very large, up to 18 × 6 cm., apex broadly
            rounded or shortly cuspidate and reflexed;
            venation prominent and coarsely reticulate on
            both surfaces . . . .   4. *E. excelsum*
        Leaflets much smaller, apex otherwise, venation not
            prominent:
            Leaflets with closely reticulate but not prominent
                venation; petiolules 1–3.5 cm. long . . . . .   5. *E. delevoyi*
            Leaflets usually with indistinct venation, subsessile or
                very shortly petiolulate . . . .   6. *E. angolense*

1. **E. cylindricum** (*Sprague*) *Sprague* in K.B. 1910: 180 (1910); Hoyle in K.B. 1932: 40(1932); Eggeling & Harris, For. Trees Brit. Emp. 4: 41, photo. 6, t. 6 (1939); C.F.A.: 310 (1951); I.T.U., ed. 2: 178, photo. 31, t. 38 (1952); F.W.T.A., ed. 2, 1: 701 (1958); Staner & Gilbert in F.C.B. 7: 184 (1958); Styles in E. Afr. Agric. Journ. 39: 411 (1974). Type: Ghana, near Mansu and Supom, *Thompson* 16 (K, holo.!)

Tree up to 55 m. or more high; bole exceptionally long and straight, scarcely tapering, free of branches for 30 m. or more; buttresses small relative to the size of the tree, seldom extending more than 3 m. up the stem; bark at first brown and smooth, becoming lighter and greyer with age, flaking off towards base in irregularly shaped plates giving a scaly appearance. Leaves up to 30 cm. long; leaflets 11–19, often alternate, mostly oblong-lanceolate or oblong-elliptic, but the proximal pairs more or less ovate, up to 12.5 × 4 cm., apex shortly acuminate, base slightly asymmetric; lateral nerves in 6–12 pairs, venation fine but prominent and closely reticulate on both surfaces; lower surface glabrous except for a few minute hairs on the midrib; petiole flattened on upper surface and slightly winged, densely puberulous. Inflorescence up to 25 cm. long. Calyx 0.1 cm. long, lobed to beyond the middle, sparsely and minutely puberulous. Petals 0.35 cm. long, sparsely and minutely puberulous. Staminal tube broadly urceolate, margin denticulate, up to 0.2 cm. long in ♂, shorter and proportionally broader in ♀; disk-ridges 20. Capsule up to 14 cm. long, cylindric, rounded or obtuse at the apex; valves 0.3 cm. thick, dehiscing first from the apex, but very soon afterwards also from the base so that the valves fall away singly, brown outside and obscurely mottled with small lenticels, cream inside with pale brown streaks; columella scarcely narrowed to the base, pale buff; scars small, inconspicuous, without hairs, connected to each other in 2 rows on each face of columella by 2 narrow slightly zigzag placental lines. Seeds (including wing) ± 7.5 × 1.8 cm., attached to columella by a lateral hilum. Fig. 15.

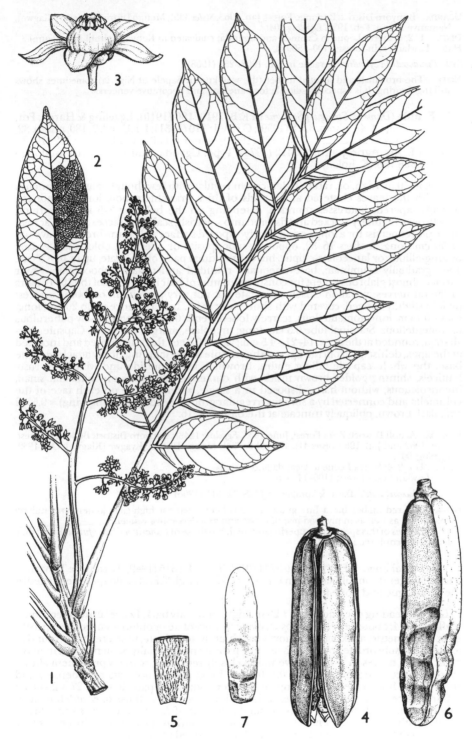

FIG. 15. *ENTANDROPHRAGMA CYLINDRICUM*—**1**, flowering shoot, × ½; **2**, leaflet, × ½; **3**, flower, × 4⁷/₁₀; **4**, capsule, × ½; **5**, inner face of valve showing distinctive markings, × ½; **6**, columella, × ½; **7**, seed, × ½. 1–3 from *Styles* 336; 4–7, from *Kennedy* 1163. Drawn by Janet Dyer.

UGANDA. Bunyoro District: Budongo Forest, Jan. 1963, *Styles* 335!; Mengo District: Kyagwe [Kiagwe], Namanve Forest, Feb. 1932, *Eggeling* 204!
DISTR. U 2, 4; Sierra Leone to Cabinda and Uganda; cultivated in Kenya (Nairobi Arboretum)
HAB. Lowland rain-forest; 1100–1500 m.

SYN. *Pseudocedrela cylindrica* Sprague in K.B. 1908: 257 (1908)

NOTE. The brownish and scented timber (the well-known Sapele of Nigeria) sometimes shows striking figuring. It is especially suitable for panelling and decorative veneers.

2. **E. utile** (*Dawe & Sprague*) *Sprague* in K.B. 1910: 180 (1910); Eggeling & Harris, For. Trees Brit. Emp. 4: 48, photo. 8, t. 8 (1939); C.F.A. 1: 310 (1951); I.T.U., ed. 2: 180, photo. 32, t. 40 (1952); F.W.T.A., ed. 2, 1: 700, t. 195 (1958); Staner & Gilbert in F.C.B. 7: 182 (1958); Styles in E. Afr. Agric. Journ. 39: 412 (1974). Type: Uganda, Bunyoro District, Budongo Forest, *Dawe* 786 (K, holo.!)

Tree to 50 m. high, or, rarely, taller; crown regular with few though massive branches and with the leaves clustered at the ends of thick branchlets; bole long, cylindrical, scarcely tapered; buttresses rounded, extending up the bole for 2–5 m.; bark thick, grey-brown, regularly cracked and fissured into squarish, scale-like pieces 3–6 cm. across; these pieces persist for a long period; newly-exposed surface pale warm brown. Leaves up to 50 cm. long; leaflets 16–32, opposite, subopposite or alternate, oblong-lanceolate, oblong-elliptic or lanceolate-elliptic, but the proximal pairs often ovate, up to 14 × 5 cm., apex gradually acuminate, base unequally rounded and usually subocordate; lower surface almost glabrous except for conspicuous tufts of hair in the axils of the 10–16 pairs of lateral nerves; venation closely reticulate but not very prominent on both surfaces; petiole slightly winged, covered with short rusty hairs. Inflorescence up to 20 cm. long. Calyx 0.1 cm. long, tomentellous, scarcely lobed. Petals 0.5 cm. long, densely puberulous to tomentellous. Staminal tube 0.3 cm. long, urceolate, margin subentire. Capsule club-shaped, rounded at the apex, 14–21 × 4.5 cm.; valves 0.8 cm. thick, thickened and incurved at the apex, dehiscing from the apex of the capsule and remaining firmly attached to the base, the whole capsule falling entire, brownish black outside and with large rusty lenticels, shining golden-brown inside with darker streaks; scars on columella small, inconspicuous, without hairs, disposed in a single, ± central line on each face of the columella and connected by a sinuous irregular groove. Seeds (including wing) ± 9.5 × 2 cm., dark brown, obliquely truncate at the base. Fig. 16.

UGANDA. Acholi District: Zoka Forest, June 1933, *Eggeling* 1243!; Bunyoro District: Budongo Forest, Sept. 1962 and Jan. 1963, *Styles* 41!, 47!, 73! & 334!; Mengo District: Kyagwe [Kiagwe], Oct. 1932, *Eggeling* 902!
DISTR. U 1, 2, 4; Sierra Leone to Uganda and Angola
HAB. Lowland rain-forest; 1100–1400 m.

SYN. *Pseudocedrela utilis* Dawe & Sprague in J.L.S. 37: 511 (1906)

NOTE. The red timber has a fine grain and has been used for high-class joinery and railway coachwork as well as in plywood manufacture and as a decorative veneer.
   The leaves of this species are superficially similar to those of *Canarium schweinfurthii* Engl.; they have been confused in herbaria.

3. **E. bussei** Engl., V.E. 3(1): 807, t. 379 (1915); T.T.C.L.: 315 (1949). Lectotype: Tanzania, 'aus der Gegend von Kilimatinde in Ugogo', t. 379 in Engl. V.E. 3(1); no specimens cited in the protologue and none found

Large spreading deciduous tree ± 20 m. high, with scaly bark. Leaves up to 30 cm. long; leaflets 10–14, opposite or subopposite, ovate to lanceolate or oblong-lanceolate, up to 9 × 4 cm., asymmetric, apex acute to shortly and bluntly acuminate, base unequally rounded to obliquely subcordate; lower surface softly pubescent, especially on nerves and in nerve axils; lateral nerves in 7–11 pairs, venation closely reticulate, scarcely prominent above, more distinct beneath: petiolules 0.1–0.5 cm. long; petiole not flattened, densely and softly hairy. Inflorescence an elongate, slender contracted panicle up to 27 × 2–4 cm.; inflorescence-axes and calyx densely pilose. Flowers similar to those of *E. utile* but much smaller; petals 0.2 cm. long. Capsule similar to that of *E. utile* but valves very closely mottled with smaller buff lenticels outside and pale buff inside with indistinct markings. Seeds (including wing) ± 9 × 2.5 cm., pale buff, rounded at the base.

TANZANIA. Shinyanga, Sept. 1935, *B.D. Burtt* 5300!; Dodoma District: below Kilimatindi, Mtewe, June 1933, *B.D. Burtt* 4449!; Iringa District: Kalonga Pass on Kilosa road, Oct. 1936, *B.D. Burtt* 6059!

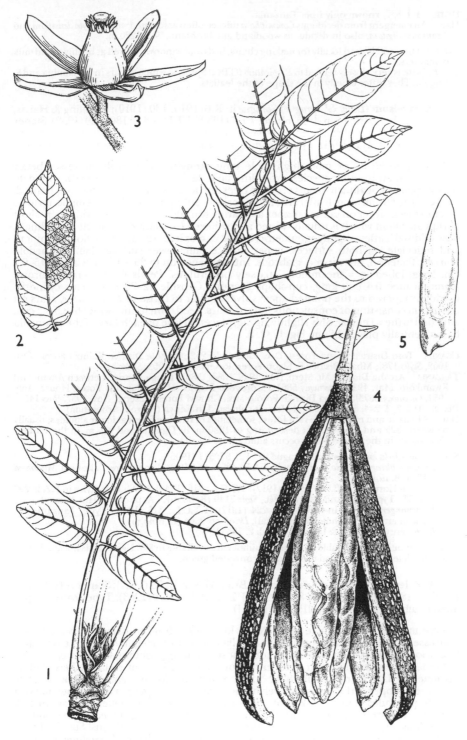

FIG. 16. *ENTANDROPHRAGMA UTILE*—**1**, leaf and apical bud, × ½; **2**, leaflet, × ½; **3**, flower, × 4½; **4**, capsule, one valve removed, × ½; **5**, seed, × ½. 1, from *Styles* 41; 2,4,5, from *Styles* 73; 3, from *Styles* 334. Drawn by Janet Dyer.

DISTR. T 1, 5, 7; known only from Tanzania
HAB. An emergent from deciduous *Commiphora* thicket, often associated with *Cordyla densiflora* and *Adansonia digitata;* also in deciduous woodland and bushland; 785–1220 m.
NOTE. The wood is used locally for making chairs, beds and windows, as well as beehives and milk containers.
  E. bussei is very closely related to *E. spicatum* (C.DC.) Sprague which occurs in N. Namibia and S. Angola. They differ chiefly in the shape of the leaflets.

4. **E. excelsum** (*Dawe & Sprague*) *Sprague* in K.B. 1910: 180 (1910); Eggeling & Harris, Trees & Timbers Brit. Emp. 4: 45, photo. 7, t. 7 (1939); I.T.U., ed. 2: 180, t. 39 (1952); Staner & Gilbert in F.C.B. 7: 188 (1958); F. White & Styles in F.Z. 2: 632 (1966); Styles in E. Afr. Agric. Journ. 39: 411 (1974). Type: Uganda, W. Ankole, 1500 m., *Dawe* 358 (K, holo.!)

Tree up to 55 m. or more high; bole long, straight, cylindrical; buttresses better developed than in other species, extending up bole for 4–5 m.; bark greyish with pale orange patches, scaling in plates on older trees and leaving shallow depressions behind. Leaves up to 60 cm. long; leaflets 10–14, opposite or subopposite, oblong-elliptic, up to 18 × 8 cm. (much larger on saplings), apex appearing rounded or retuse, but with a very short abrupt apiculum which is usually twisted and reflexed (sometimes subacute in saplings), base of distal leaflets slightly asymmetric; lateral nerves in 8–12 pairs, venation prominent and reticulate on both surfaces; upper surface dull; lower surface glabrous; petiole scarcely flattened, not winged, glabrous. Inflorescence up to 25 × 15 cm. Calyx 0.15–0.2 cm. long, lobed to ± the middle, sparsely puberulous. Petals 0.6–0.7 cm., glabrous. Staminal tube 0.4 cm. long, cup-shaped, entire. Capsule 12–20 × 3–4 cm., cylindric, gradually tapered to the pointed apex; valves similar to those of *E. angolense*; scars on columella consisting of conspicuous tufts of hairs. Seeds (including wing) 6–8.5 × 2 cm., attached to the ± central, only slightly zigzag, placental line of each face of the columella by a ± centrally placed hilum. Fig. 17/1, 2.

UGANDA. Toro District: Itwara Forest, Dec. 1962, *Styles* 252!; Ankole District: Kalinzu Forest, Oct. 1962, *Styles* 178!; Mbale District: S. Elgon Forest Reserve, Jan. 1963, *Styles* 319!
TANZANIA. Arusha District: Mt. Meru, Nov. 1949, *Hoyle* 1388!; Tanga District: between Amani and Kwamkoro, Mar. 1939, *Greenway* 5861!; Morogoro District: above Bunduki Forest House, Aug. 1951, *Greenway* 8673!; Mbeya District: Poroto Mts., Chuvwi Forest Reserve, Nov. 1937, *Ross* 14!
DISTR. U 2, 3; T 2–4, 6–8; Zaire and Malawi; cultivated in Kenya (Nairobi Arboretum)
HAB. Montane and mid-altitude rain-forest, sometimes in riverine forest at lower altitudes; usually between 1525 and 2150 m., but descending to 1280 m. in Uganda and 925 m. in the Eastern Usambaras; in the Poroto Mts. it occurs almost pure locally
SYN. *Pseudocedrela excelsa* Dawe & Sprague in J.L.S. 37: 511 (1906)
  Entandrophragma speciosum Harms in Z.Ä.E.: 429, t. 48 (1912). Type: Zaire, Lake Kivu, *Mildbraed* 1203 (B, holo.†)
  E. stolzii Harms in N.B.G.B. 7: 224 (1917); T.T.C.L.: 316 (1949); F. White & Styles in F.Z. 2: 292 (1963). Type: Tanzania, Rungwe Mt., *Stolz* 2149 (B, holo.†, EA, photo.!, FHO, K, iso.!)
  E. deiningeri Harms in N.B.G.B. 7: 224 (1917); T.T.C.L.: 316 (1949). Type: Tanzania, W. Usambara Mts., Lushoto [Wilhelmstal], *Deininger* 2964 (B, holo.†)
  E. sp. sensu Eggeling & Dale, I.T.U., ed. 2: 183 (1952)
NOTE. Although considerable quantities of timber are available in SW. Uganda it is not much used because it tends to warp and twist badly if converted green.

5. **E. delevoyi** *De Wild.* in Ann. Soc. Sci. Brux. 47, sér. B: 78 (1927); Staner & Gilbert in F.C.B. 7: 181 (1958); F. White & Styles in F.Z. 2: 292, t. 55A (1963). Type: Zaire, Upper Shaba, Lubumbashi, *Delevoy* 915 (BR, holo.!)

Tree up to 35 m. tall; bole up to 20 m. long and 1.5 m. in diameter, very slightly buttressed at the base; bark grey-brown, exfoliating in large irregular pieces. Leaves up to 25 cm. long; leaflets 6–10, opposite or subopposite, oblong to oblong-lanceolate, up to 9.5 × 4 cm., apex suddenly contracted in a short acumen, base of distal leaflets slightly asymmetric; lateral nerves in 8–12 pairs, venation closely reticulate but scarcely prominent; upper surface glossy; lower surface glabrous except for a few minute hairs on midrib; petiolules slender, 1–2 (3.5) cm. long; petiole scarcely flattened, not winged, glabrous. Inflorescence up to 15 × 8 cm. Calyx 0.1 cm. long, lobed to the middle, puberulous, especially on the margin. Petals 0.5–0.7 cm. long, glabrous. Staminal tube 0.3–0.4 cm. long, cup-shaped, subentire. Capsule similar to that of *E. excelsum* but more suddenly contracted to the shortly pointed apex, and with smaller, much less conspicuous lenticels. Fig. 17/3, 4.

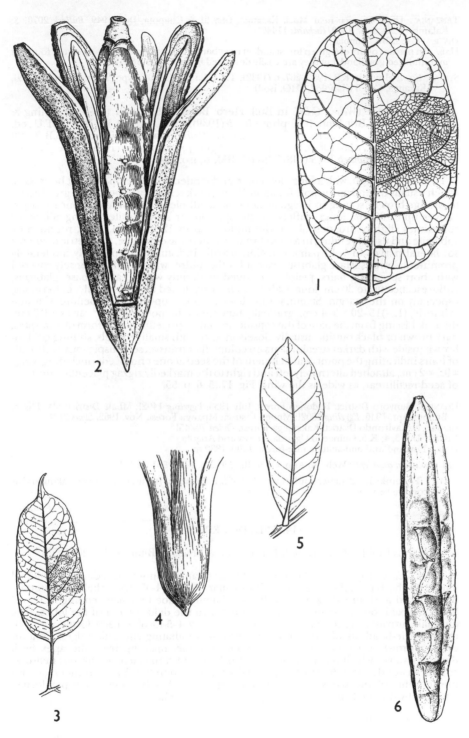

FIG. 17. *ENTANDROPHRAGMA EXCELSUM*—**1**, leaflet, × ½; **2**, capsule, one valve removed, × ½.
E. DELEVOYI—**3**, leaflet, × ½; **4**, apex of capsule, × ½. *E. ANGOLENSE*—**5**, leaflet, × ½; **6**, columella, × ½.
1, from *Styles* 319; 2, from *Styles* 252; 3, from *Lawton* 644; 4, from *Angus* 374; 5, from *Styles* 202; 6, from *Styles*
215. Drawn by Janet Dyer.

TANZANIA. Ufipa District: near Muzi, Kasanga, *Grey* 5! & Chapota, Dec. 1949, *Bullock* 2070! & Kalambo Falls, Sept. 1959, *Richards* 11449!
DISTR.   **T** 4; Zaire, Zambia
HAB.   In patches of dry evergreen forest and on river banks; relict trees are also sometimes found in man-induced grassland, but they are easily destroyed by fire; 1200–1675 m.
SYN.   *E. lucens* Hoyle in K.B. 1932: 267, t. (1932); T.T.C.L.: 316 (1949). Type: Tanzania, near Muzi, Kasanga, *Grey* 5 (K, holo.!, FHO, iso.!)

6. **E. angolense** (*Welw.*) *C.DC.* in Bull. Herb. Boiss. 2: 582, t. 21 (1894); Eggeling & Harris, For. Trees Brit. Emp. 4: 36, photo. 5, t. 5 (1939); C.F.A. 1: 308, t. 14 (1951); I.T.U., ed. 2: 176, photo. 30, t. 37 (1952); F.W.T.A., ed. 2, 1: 700 (1958); Staner & Gilbert in F.C.B. 7: 190 (1958); K.T.S.: 267, t. 55 (1961); Styles in E. Afr. Agric. Journ. 39: 410 (1974). Type: Angola, Golungo Alto, *Welwitsch* 1313 (LISU, holo.!, BM, K, iso.!)

Tree to 50 m. high, with a long clean bole and moderately developed blunt buttresses, which usually extend up the bole for about 2.5 m.; bark relatively smooth, pale grey-brown to orange-brown, scaling in irregular large or small pieces which leave concave scars. Leaves up to 30 cm. long; leaflets 14–20, opposite or subopposite, oblong-elliptic or sometimes slightly broadest above the middle, up to 12.5 × 5 cm., apex rounded or subacute and usually ending in a short broad acumen; base cuneate, symmetric or almost so; lateral nerves in 7–12 pairs, venation usually indistinct, exceptionally moderately prominent; lower surface glabrous except for the midrib which is often densely covered with short spreading hairs; petiole unwinged or slightly winged at the base, glabrous. Inflorescence up to 30 cm. long. Calyx 0.1 cm. long, lobed to ± the middle, puberulous, especially on the margin. Staminal tube 0.4 cm. long, cup-shaped, subentire. Capsule cylindric, (12–)15–20 × 3–4 cm., gradually tapered to the pointed apex; valves ± 0.3 cm. thick, dehiscing from the base of the capsule and falling together in the form of a calyptra, dark brown or black outside, usually closely mottled with small lenticels, shining golden-brown inside with darker streaks; scars on columella ± transverse, conspicuous, with a tuft of hairs indicating the point of attachment of the seed at one end. Seeds (including wing) ± 9.5 × 2 cm., attached alternately left and right to the markedly zigzag placental line; base of seed rectilinear, as wide as the wing. Fig. 17/5, 6, p. 55.

UGANDA.   Bunyoro District: Budongo Forest, July 1933, *Eggeling* 1402!; Mbale District: SW. Elgon, Bupoto, Apr. 1946, *Eggeling* 5602!; Mengo District: Mpanga Forest, Nov. 1962, *Styles* 213!
KENYA.   N. Kavirondo District: Kakamega Forest, *Faden* 70/45!
DISTR.   **U** 2, 3, 4; **K** 5; Guinée to Uganda, Kenya and Angola
HAB.   Lowland and mid-altitude rain-forest; 1100–1830 m.

SYN.   *Swietenia angolensis* Welw. in Ann. Conselho Ultram. 1858: 561, 587 (1859)

NOTE.   The pinkish red timber is inferior to that of *E. cylindricum*. It is used in cabinet making and as a decorative veneer.

14. PSEUDOCEDRELA

Harms in E.J. 22: 153 (1895); Pennington & Styles in Blumea 22: 520 (1975)

Trees. Leaves paripinnate. Flowers unisexual (plants monoecious). Calyx (4–)5-lobed almost to the base, lobes ovate or suborbicular. Petals (4–)5, free, slightly contorted, boat-shaped and spreading in open flower. Staminal tube urceolate, ending in (8–)10 bifid, reflexed lobes, the anthers inserted between the deltate teeth of the lobes. Disk annular, surrounding the base of the ovary. Ovary 4–5-locular, each locule with 4–6 ovules; style-head discoid, upper surface with 4–5 radiating stigmatic ridges. Fruit an erect, narrowly claviform, woody, septifragal capsule, opening from the apex by 5 divergent valves which remain attached at the base and connected by a fibrous network; columella woody, extending to the apex of the capsule, sharply 4–5-angled, indented with the imprints of the seeds; seed-scars inconspicuous. Seeds winged, 4–5 per locule, attached by the seed-end to the distal part of the columella.

One species in tropical Africa.

P. kotschyi (*Schweinf.*) *Harms* in E.J. 22: 154 (1895); Sprague in K.B.: 181 (1910); V.E. 3(1): 803, fig. 378 (1915); Chalk et al., For. Trees Brit. Emp. 2: 66, t. 13 & photo. (1933); Pellegrin in Not. Syst. 9: 34, fig. 2 (1940); F.P.S. 2: 331, fig. 122 (1952); I.T.U., ed. 2: 193, fig. 45 (1952);

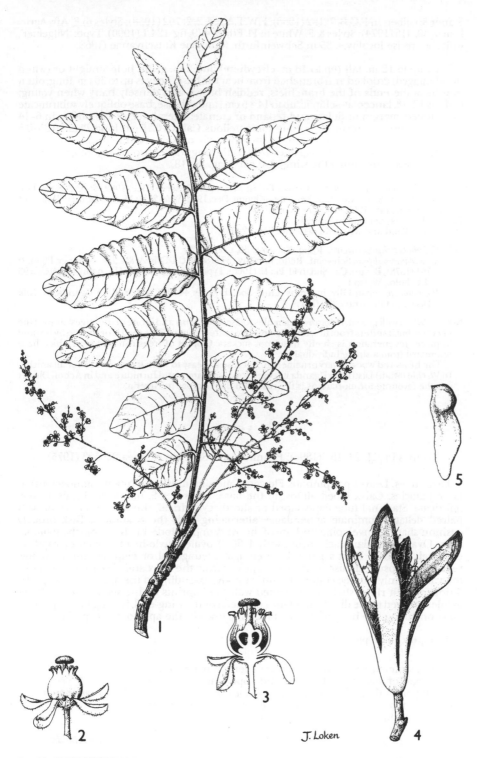

FIG. 18. *PSEUDOCEDRELA KOTSCHYI*—**1**, flowering shoot, × ½; **2**, ♂ flower, × 2⁷⁄₁₀; **3**, half ♂ flower, vertical section, × 2⁷⁄₁₀; **4**, capsule, × ½; **5**, seed, × ½. 1, 4, 5, from *Eggeling* 1502; 2, 3, from *Styles* 278. Drawn by Julia Loken.

Staner & Gilbert in F.C.B. 7: 192 (1958); F.W.T.A., ed. 2, 2: 702 (1958); Styles in E. Afr. Agric. Journ. 39: 416 (1974); Styles & F. White in Fl. Eth. 3: 479, fig. 124.1 (1990). Type: 'Nilgebiet', without precise locality, t. 35 in Schweinfurth, Reliquiae Kotschyanae (1868)

Tree up to 12 m. tall (up to 18 m. elsewhere) and 2 m. girth; bole straight or (when fire-damaged) crooked and branched from near the base. Leaves up to 30 cm. long, often in tufts at the ends of the branchlets, reddish brown and densely hairy when young; leaflets 12–18, lanceolate-elliptic, up to 14 × 5 cm., apex obtuse, base obliquely subtruncate or rounded, margin undulate and repand or crenate; lateral nerves prominent in 6–14 pairs. Inflorescence up to 28 cm. long, tomentellous. Calyx ± 0.15 cm. long. Petals 0.35–0.5 cm. long, with scattered hairs on the median axis outside. Staminal tube (including appendages) ± 0.3 cm. long. Capsules 7–14.5 cm. long; valves brown, smooth, with paler lenticels. Seeds (including the wing) 4–6 cm. long. Fig 18.

UGANDA. W. Nile District: N. Era Crown Forest Protection Reserve, Dec. 1962, *Styles* 278!; Acholi
    District: 1.5 km. S. of Aswa R., 32 km. Gulu–Kitgum, Dec. 1962, *Styles* 258!; Mengo District: 7 km. E. of
    Nakasongola, Jan. 1956, *Langdale-Brown* 1863!
DISTR. U 1, 3, 4; Senegal to Ethiopia
HAB. Woodland and wooded grassland; 750–1200 m.

SYN. *Cedrela kotschyi* Schweinf., Reliq. Kotschy.: 36, t. 35 (1868)
    Soymida roupalifolia Schweinf., Reliq. Kotschy.: 37 (1868); C.DC. in A. & C.DC., Monogr. Phan. 1:
        751 (1878); Broun, Cat. Sudan Fl. Pl.: 15 (1906). Type: Sudan, Fazughli [Fesoglu], *Cienkowski* 93
        (LE, holo., W, iso.)
    Boswellia sp. sensu Oliv. in Trans. Linn. Soc., Bot. 29: 44 (1872); based on Uganda, W. Nile
        District, Madi, *Grant* 739 (K!)

NOTE.   When well grown, *P. kotschyi* resembles *Khaya senegalensis* and is one of the largest trees of the
    savanna, but it suffers from fire, and distorted trees are frequent. The seed is destroyed by fire and
    in places regeneration is chiefly from root suckers. Clumps of trees can be found which have
    originated from a single individual.
        The heavy red wood is very ornamental. Because it is easy to work it is esteemed for all articles.
    In W. Nile (Madi) District in Uganda it is used for door frames and furniture and in Acholi District
    it is the favourite for mortars, as is the case with the Yorubas in Nigeria.

15. **LOVOA**

Harms in E.J. 23: 164 (1896); Pennington & Styles in Blumea 22: 523 (1975)

Large trees. Leaves paripinnate. Flowers 4-merous, unisexual (plants monoecious) in large panicles. Calyx lobed almost to the base; lobes 2 + 2, imbricate. Petals 4, free, imbricate. Staminal tube cup-shaped or shortly cylindrical, the margin entire or with paired deltate-acuminate appendages alternating with the 8 anthers. Disk broadly cushion-shaped, enveloping the base of the ovary or pistillode, but free from the staminal tube. Ovary 4-locular, each locule with 4–6(–8) ovules; style-head discoid or capitate, obscurely 4-lobed. Fruit a pendulous, elongate, tetragonal or ellipsoid, thinly woody, septifragal capsule, dehiscing from the apex or from the apex and base simultaneously, the valves thinly woody; columella softly woody, extending to the apex of the capsule, 4-ridged, each ridge shallowly indented with the imprints of 1–2 seeds. Seeds ± 2 per locule, attached to the distal part of the columella by the wing-end, leaving inconspicuous scars on falling, the body of the seed hanging towards the apex of the capsule.

Two species in tropical Africa.

Leaflets asymmetrical; inflorescence-axes densely puberulous;
    margin of staminal tube with bifid appendages; capsule
    ellipsoid, valves separating from the apex first   .   .   .   1. *L. swynnertonii*
Leaflets symmetrical; inflorescence-axes glabrous; margin of
    staminal tube entire; capsule tetragonal, valves separating
    from the base first or from the apex and base
    simultaneously   .   .   .   .   2. *L. trichilioides*

1. **L. swynnertonii** *Bak.f.* in J.L.S. 40: 41, t. 3 (1911); T.T.C.L.: 317 (1949), pro parte excl. specim. *Gillman* 285; I.T.U., ed. 2: 193 (1952); Staner & Gilbert in F.C.B. 7: 195 (1958); K.T.S.: 269 (1961); F. White & Styles in F.Z. 2: 293, t. 56 (1963); Styles in E. Afr. Agric. Journ. 39: 416 (1974). Type: Zimbabwe, Chirinda, *Swynnerton* 16 (BM, holo.!, K, iso.!)

Evergreen tree occasionally reaching a height of 50 m.; bole fluted or slightly buttressed at the base to a height of 2 m., long and straight, sometimes 30 m. to first branch, slender, up to 2 m. in diameter. Leaves up to 30 cm. long, pubescent when young; leaflets usually 10–16, ± oblong-elliptic or lanceolate-elliptic, slightly falcate, up to 10 × 4 cm., apex shortly acuminate, base markedly asymmetric; lateral nerves in ± 16 closely spaced pairs; petiole flattened. Inflorescence an axillary panicle up to 10 cm. long. Calyx 0.1 cm. long, puberulous especially on the margins. Petals 0.25–0.3 cm. long, glabrous. Staminal tube (including appendages) 0.2–0.3 cm. long. Capsule up to 5.5 × 2 cm.; valves brownish black, with scattered, minute, white lenticels, separating first from the apex and remaining attached for some time before falling. Seeds (including wing) up to 4.5 × 1 cm.

UGANDA. Toro District: Kibale Forest Reserve, Nov. 1962, *Styles* 245a!; Ankole District: Kalinzu Forest Reserve, June 1938 *Eggeling* 3730!; Mengo District: Lwamunda Forest Reserve, Nov. 1962, *Styles* 210!
KENYA. Meru District: Meru, July 1938, *Leakey* 1241! & Thura Forest, *Porter* 915!; Kwale District: Mrima Hill, Jan. 1964, *Verdcourt* 3936a! & Dec. 1976, *White* 11334!
TANZANIA. Moshi District: Rau Forest Reserve, *Baldock*! & Oct. 1935, *Bancroft* 892!; Morogoro District: Mtibwa Forest Reserve, Mar. 1939, *Wigg* 1525!
DISTR. U 2, 4; K 4, 7; T 2, 6; also in E. Zaire, Zimbabwe and Mozambique
HAB. Lowland and mid-altitude rain-forest; 180–1525 m.

NOTE. This tree produces a beautiful furniture-timber which is dark brownish red, but is frequently cross-grained and difficult to work. Because it is rather rare, except in Toro District, it is generally marketed with *L. trichilioides*. In Zimbabwe it was formerly used for outdoor work, being very durable and untouched by insects, but is now so rare that without protection, it is unlikely to survive.

2. **L. trichilioides** *Harms* in E.J. 23: 165 (1896); C.F.A. 1: 311 (1951); F.W.T.A., ed. 2, 1: 702 (1958); Staner & Gilbert in F.C.B. 7: 194 (1958); Styles in E. Afr. Agric. Journ. 39: 415 (1974). Type: Angola, Lunda, R. Lovo, *Marques* 232 (COI, holo., LISU, iso.)

Evergreen tree up to 40 m. tall; bole slender, cylindrical; buttresses short, indistinct, rarely more than 2 m. high. Leaves up to 24 cm. long, glabrous; leaflets usually 10–14, ± elliptic, up to 16 × 4 cm., apex shortly acuminate, base cuneate; lateral nerves in ± 20 closely spaced pairs; petiole flattened and slightly winged. Inflorescence a robust, axillary or terminal panicle up to 30 cm. long. Calyx 0.1 cm. long, glabrous except for a ciliate margin. Petals 0.4–0.65 cm. long, glabrous. Staminal tube 0.2–0.3 cm. long. Capsule up to 6 × 1.3 cm.; valves black separating first from the base, or from the apex and base simultaneously. Seeds (including wing) up to 5 × 1 cm. Fig. 19, p. 60.

UGANDA. Bunyoro District: Budongo Forest, Oct. 1932, *Harris* 1107!; Mengo Distict: Lwamunda Forest Reserve, Nov. 1962, *Styles* 205!; Masaka District: Wabitembe Forest Reserve, Nov. 1962, *Styles* 198!
TANZANIA. Bukoba District: Rubare Forest Reserve, Nov. 1933, *Pitt-Schenkel* 929!
DISTR. U 2, 4; T 1; in th canopy of Guineo-Congolian rain-forest and related types from Sierra Leone to Uganda and Tanzania and southwards to Angola
HAB. Rain-forest; in Uganda it is extremely common in the Lake Shore forests but rare elsewhere; regeneration is abundant and the tree is increasing in number in young forests; 1100–1200 m.

SYN. *L. brachysiphon* Sprague in J.L.S. 37: 508 (1906). Type: Uganda, Toro District, 1500 m., *Dawe* 457 (K, holo.!)
    L. budongensis Sprague in J.L.S. 37: 508 (1906). Type: Uganda, Bunyoro District, Budongo Forest, *Dawe* 808 (K, holo.!)
    L. brownii Sprague in J.L.S. 37: 509 (1906); Eggeling & Harris, For. Trees Brit. Emp. 4: 76, photo. 12, t. 12 (1939); T.T.C.L.: 317 (1949); I.T.U., ed. 2: 191, fig. 44 (1952). Type: Uganda, Entebbe District, *E. Brown* 243 (K, holo.!, FHO, iso.!)
    [*L. swynnertonii* sensu T.T.C.L.: 317 (1949), pro parte quoad specim. *Gillman* 285, non Bak.f.]

NOTE. This is a very important timber-producing species in Uganda and is heavily exploited. Although boles do not reach any very great girth, they are usually long with a high quantity of convertible timber, which is hard and reddish, works well and is mostly used for furniture.

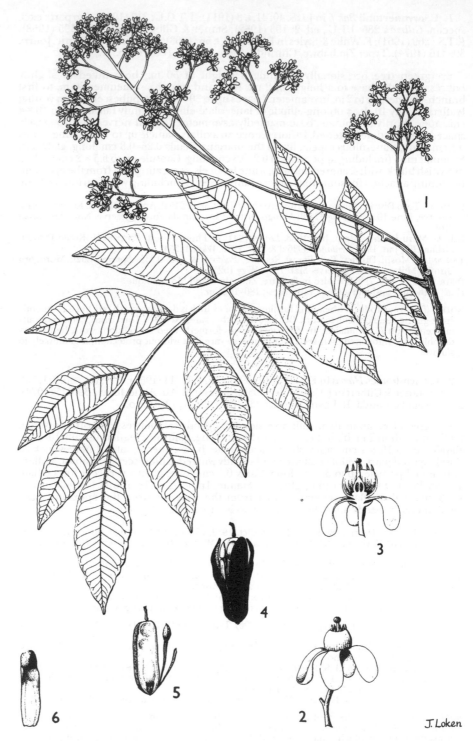

J. Loken

FIG. 19. *LOVOA TRICHILIOIDES*—1, flowering shoot, × ½; 2, ♂ flower, × 1⅘; 3, half ♂ flower, vertical section, × 1⅘; 4, capsule, × ½; 5, columella, showing attachment of one seed and base of one valve, × ½; 6, seed, × ½. 1, from *Gibson* 8; 2,3, from *Styles* 14; 4–6, from *Styles* 15. Drawn by Julia Loken.

16. **CARAPA**

Aubl., Hist. Pl. Guiane. Fr. 2, Suppl.: 32, t. 387 (1775); Noamesi, A revision of the *Xylocarpeae*, unpublished thesis University of Wisconsin: 35 (1958); Pennington & Styles in Blumea 22: 524 (1975)

Trees of very variable habit and size. Leaves usually paripinnate and with an apical gland, or exceptionally imparipinnate, often crowded at the ends of stout branchlets. Flowers 4–5(–6)-merous, unisexual, in large much-branched panicles. Calyx small, lobed almost to the base; lobes imbricate. Petals 4–5(–6), slightly contorted, spreading in open flowers. Staminal tube urceolate, very similar to that of *Khaya*. Disk cushion-shaped, surrounding the base of the ovary and free from the staminal tube. Ovary 4–5(–6)-locular, with 2–8 ovules in each locule; style-head discoid with a crenulate margin. Fruit a large, pendulous, leathery, subglobose septifragal capsule, which falls from the tree entire and breaks open on striking the ground; columella poorly developed. Seeds large, usually about 12–20, subangular, with a woody but buoyant outer covering.

2 species; *C. procera* is widespread in tropical Africa and has a restricted distribution in South America; it is closely related to, and may prove to be conspecific with, the widely distributed neotropical *C. guianensis* Aubl. (see F. White in Bothalia 14: 400 (1983)).

C. procera *DC.*, Prodr. 1: 626 (1824); C.DC. in A. & C. DC., Monogr. Phan. 1: 716 (1878); Harms in Z.A.E.: 433 (1912); Pellegrin in Not. Syst. 9: 27 (1940); C.F.A. 1: 311 (1951); Staner & Gilbert in F.C.B. 7: 197 (1958); F.W.T.A., ed. 2, 1: 702 (1958); Styles in Pennington, Fl. Neotrop. Monogr. 28: 414, fig. 84, map 86 (1981). Type: origin and collector uncertain; possibly *Forsyth*, possibly collected in W. Africa (G-DEL, iso.!)

Badly shapen understorey tree (in E. Africa) up to 25 m. tall, but usually smaller (elsewhere sometimes erect and up to 30 m. or more in height); branches widespreading and arching; bole unbuttressed but often fluted and branched low down. Leaves up to 1.5 m. long, mostly crowded at the ends of stout branchlets, bright red when young; leaflets 6–18 or more, usually oblong-elliptic or oblanceolate-elliptic, up to 40 × 16 cm., apex rounded and usually ending in an abrupt very short glandular acumen, sometimes more distinctly acuminate (especially juvenile leaves), base narrowly or broadly cuneate, slightly asymmetric; lateral nerves prominent, in ± 10 widely spaced pairs; lower surface glabrous, with a network of impressed veins. Inflorescence a pyramidal panicle up to 70 cm. or more long. Calyx lobed, ± 0.2 cm. long, glabrous, except for the ciliolate margin. Petals up to 0.9 cm. long, glabrous except for marginal cilia. Staminal tube up to 0.8 cm. long. Capsules 1–3 per infructescence, subglobose, umbonate or rostrate, 12–15 cm. in diameter. Seeds (in E. Africa) usually 10–20 per capsule, ± 3 cm. long, dark brown, shining. Fig. 20, p.62.

UGANDA. Toro District: Kibale Forest Reserve, Nov. 1962, *Styles* 245!; Ankole District: Kalinzu Forest Reserve, Oct. 1962, *Styles* 177!; Mengo District: Entebbe, Nov. 1932, *Eggeling* 710!
TANZANIA. Bukoba District: Kiamawa, *Gillman* 421!; Buha District: Mukugwa [Mukugwe] R., 48 km. S. of Kibondo, July 1951, *Eggeling* 6204!
DISTR. U 2, 4; T 1, 4; from Senegal, throughout the Guineo-Congolian rain-forest (and in similar forests beyond) to Angola and East Africa; also on the island of S. Tomé and in tropical America
HAB. In Lake-shore, riparian and mid-altitude forest (especially where drainage is impeded); 1140–1830 m.
SYN. *C. grandiflora* Sprague in J.L.S. 37: 507 (1906); Harms in Z.A.E.: 433 (1912); Eggeling & Harris, For. Trees Brit. Emp. 4: 29, fig. 4, photo. 4 (1939); T.T.C.L.: 314 (1949); I.T.U., ed. 2: 172, fig. 36 (1952); Staner & Gilbert in F.C.B. 7: 196 (1958); F.W.T.A., ed. 2, 1: 702 (1958); Styles in E. Afr. Agric. Journ. 39: 409 (1974). Type: Uganda, W. Ankole Forest, *Dawe* 351 (K, holo.!)
NOTE. The reddish brown timber, known as Uganda Crabwood, resembles true mahogany (*Swietenia*) and is used for furniture and interior fittings. Although it is locally plentiful, the small size of the bole and the frequent fluting at its base reduce the yield to only 1–2 small logs per tree. Extraction costs are therefore expensive.
     Carapa in Africa is extremely variable in habit, leaf length, leaflet shape and number, the branching of the inflorescence, flower size, ovule number, and the size of the fruit and the degree to which resinous warts are developed on its surface. Some of this variation is correlated with ecology and geography but only weakly so, and there is no convincing evidence (at least in the herbarium) that there is more than one African species.
     C. grandiflora was thought to differ from *C. procera* in having larger flowers and smaller, smooth fruits containing fewer seeds, but these characters have been found to be unreliable. Much of the E. African material of *C. grandiflora* is indistinguishable from that of *C. procera* from West and Central Africa.

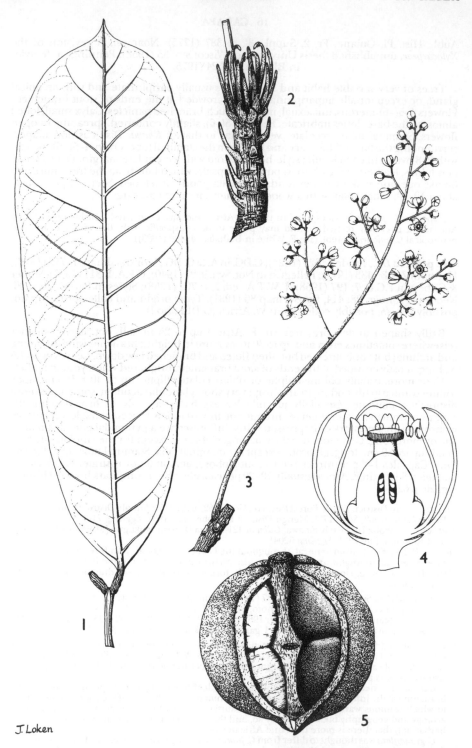

J. Loken

FIG. 20. *CARAPA PROCERA*—1, leaflet, × ⁹⁄₁₀; 2, apical bud, × ⁹⁄₁₀; 3, inflorescence, × ⁹⁄₁₀; 4, half of ♀ flower, vertical section, × ⁹⁄₁₀; 5, capsule, one valve removed, × 4⁹⁄₁₀. 1, from *Styles* 245; 2, from *Gutzwiller* 3323; 3, from *Styles* 157; 4, from *Styles* 177; 5, from *Styles* 20. Drawn by Julia Loken.

17. **XYLOCARPUS**

Koenig in Naturforscher 20: 2 (1784); Noamesi, A revision of the *Xylocarpeae*, unpublished
thesis Univ. of Wisconsin: 103 (1958); Pennington & Styles in Blumea 22: 525 (1975)

Trees. Leaves paripinnate. Flowers unisexual, in short little-branched axillary panicles.
Calyx 4-lobed to ± the middle, valvate. Petals 4, contorted, spreading or reflexed. Staminal
tube urceolate, similar to that of *Khaya*. Disk large, red, cushion-shaped, surrounding the
base of the ovary and free from the staminal tube. Ovary 4-locular, with 3–6 ovules in each
locule; style-head discoid, margin crenellate, upper surface with 4 radiating stigmatic
grooves. Fruit a large, pendulous, leathery, subglobose, tardily dehiscent septifragal
capsule; columella poorly developed. Seeds 8–20, large, angular, outer surface somewhat
rounded, with a corky outer covering.

2 or 3 species in coastal habitats from East Africa to the Pacific.

Noamesi (loc. cit.) recognizes three species of *Xylocarpus* and records them all from East Africa. In
our opinion (F.Z. 2: 295 (1963)) the two specimens of *X. mekongensis* Pierre that he cites from Kenya
and Mozambique belong to *X. granatum*. On the coast of E. Africa the taxonomy of *Xylocarpus* is
clear-cut. In Asia and the Pacific, however, the taxonomy remains unresolved and needs critical
study. The nomenclature is also confused. We think that Noamesi's interpretation of *X. granatum*
and *X. moluccensis* should be upheld. Mabberley, however, (Malaysian Forester, 45: 448 (1982)) in a
precursory note for the 'Tree Flora of Malaya' (vol. 4: 258 (1989)) has suggested otherwise. He has (in
our opinion prematurely) substituted the name *X. rumphii* (Kostel.) Mabb. (*Carapa rumphii* Kostel.,
1836) for *X. moluccensis* sensu Robinson (printed ticket on specimen, *Pl. Rumph. Amb.* No. 491
collected on 19 September 1913), Merrill ('An interpretation of Rumphius's Herbarium
Amboinense': 307 (1917)), Noamesi, F. White & Styles and other authors. Mabberley uses the name
X. moluccensis (Lam.) M. Roemer for the plant that Noamesi and others (e.g. Tomlinson, 'The botany
of mangroves': 281 (1986)) call *X. mekongensis*.

Leaflets elliptic, oblong-elliptic or obovate-elliptic, apex
    rounded, obtuse or emerginate; fruit up to 20 cm. in
    diameter; bark smooth, flaking; surface roots forming
    ribbon-like pneumatophores . . . .   1. *X. granatum*
Leaflets ovate or ovate-lanceolate, apex subacuminate; fruit up
    to 8 cm. in diameter; bark rough and longitudinally
    fissured; without ribbon-like pneumatophores. . . .   2. *X. moluccensis*

1. **X. granatum** *Koen.* in Naturforscher 20: 2 (1784); A. Juss. in Mém. Mus. Nat. Hist.
Nat., Paris 19: 244 (1830); Parkinson in Ind. For. 60: 138, t. 15 (1934); Merr., Interpret.
Rumph. Herb. Amboina: 306 (1917); Ridley in K.B. 1938: 288 (1938); Noamesi, A revision
of the *Xylocarpeae*: 107 (1958); F. White & Styles in F.Z. 2: 295, t. 57A (1963); Tomlinson,
The botany of mangroves: 278, figs. B39 & 42–43. Neotype not yet chosen.

Tree 3–15 m. tall; bark smooth, pale green or yellowish brown and peeling in irregular
patches so that the trunk is blotchy. Surface roots laterally compressed and forming a
spreading network of ribbon-like pneumatophores with the upper edges protruding
above the mud and suggesting a mass of snakes; peg-like pneumatophores have not been
recorded. Leaves up to 10 cm. long, glabrous, drying reddish brown; leaflets 2–6,
coriaceous, elliptic, oblong-elliptic or obovate-elliptic, up to 12 × 15 cm., apex usually
rounded, rarely obtuse or emerginate, base cuneate. Inflorescence (2–)4–7 cm. long.
Calyx up to 0.3 cm. long, glabrous. Petals up to 0.65 cm. long, glabrous. Staminal tube up to
0.5 cm. long, glabrous. Fruit subglobose and obscurely 4-sulcate, usually 14–20 cm. in
diameter, a leathery, tardily dehiscent, septifragal capsule which liberates the seeds on
falling to the ground. Seeds 4–8 cm. long, germinating in the capsule or soon after their
release. Fig. 21/1–5.

KENYA. Kwale District: Vanga, Jan. 1930, *R.M. Graham* in *F.D.* 2236; Mombasa, Tudor House Beach,
    Aug. 1965, *Sangai* 846!; Lamu District: Manda I., Takwa Creek, Feb. 1956, *Greenway & Rawlins* 8869!
TANZANIA.   Tanga District: Bomandani, Aug. 1953, *Drummond & Hemsley* 3663!; Rufiji District: Mafia
    I., Ras Mbisi–Mfuruni, Oct. 1937, *Greenway* 5373!; Zanzibar I., Mazizini [Massazini], June 1959,
    Faulkner 2286!; Pemba I., Makongwe I., Dec. 1930, *Greenway* 2732!
DISTR.   K 7; **T** 3, 6; **Z**; **P**; Somalia, Mozambique, Aldabra, Madagascar and throughout most of the
    Old World tropics to Australia, Fiji and Tonga
HAB.   In tidal mud of mangrove swamps, especially towards their upper limits

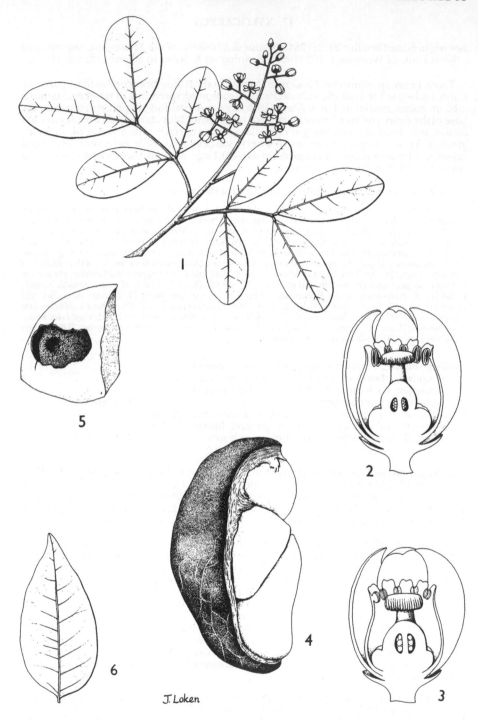

J. Loken

FIG. 21. *XYLOCARPUS GRANATUM*—**1**, flowering shoot, × ⁹/₁₀; **2**, half ♂ flower, vertical section, × 4⁷/₁₀; **3**, half ♀ flower, vertical section, × 4⁷/₁₀; **4**, part of fruit, × ⁹/₁₀; **5**, seed with part of corky aril removed, × ⁹/₁₀. *X. MOLUCCENSIS*—**6**, leaflet, × ⁹/₁₀. 1, from *Graham* 368; 2, from *Faulkner* 1512; 3, from *Hoogland* 4302; 4, from *Bond* s.n.; 5, from *Harris* s.n.; 6, from *Greenway* 5058. Drawn by Julia Loken.

Syn. *Carapa obovata* Blume, Bijdr.: 179 (1825); Baillon in Grandid., Hist. Nat. Pl. Madag. 3, t. 260
(1893); T.S.K., ed. 2: 103 (1936): Type: Java, *Blume* 1620 (L, holo., U, iso.)
   Xylocarpus benadirensis Mattei in Boll. Ort. Bot. Palermo 7: 99 (1908); Ridley in K.B. 1938: 288
(1938), descr. ampl.; T.T.C.L.: 322 (1949); K.T.S.: 276 (1961). Type: Somalia, *Macaluso* (PAL,
holo.)

Note.   In the Far East, the mahogany-like timber is valued for furniture and carpentry, but in Africa
it is of too small a size and too scarce to be of much commercial importance; it is sometimes used
for the masts of dhows. On Mafia I. a decoction of the crushed fruits is drunk as an aphrodisiac
(*Greenway* 5373).

2. **X. moluccensis** (*Lam.*) *M. Roemer*, Synops. Monogr. Hesper.: 124 (1846); Parkinson
in Ind. For. 60: 142, t. 16 (1934); Merr., Interpret. Rumph. Herb. Amboina: 307 (1917);
Ridley in K.B. 1938: 291 (1938); T.T.C.L.: 322 (1949); K.T.S.: 276 (1961); F. White & Styles in
F.Z. 2: 297, t. 57B (1963). Lectotype still to be chosen

Tree 3–15 m. tall; bark rough. Ribbon-like pneumatophores absent. Leaves up to 16 cm.
long, glabrous drying yellowish green; leaflets 2–6, subcoriaceous, ovate or ovate-
lanceolate, up to 12 × 6 cm., tapering from near the asymmetric obtuse or subtruncate
base to the subacuminate apex. Inflorescence 5–15 cm. long, when well developed a lax
raceme of lax cymes. Calyx ± 0.2 cm. long, glabrous. Petals up to 0.7 cm. long, glabrous.
Staminal tube up to 0.55 cm. long, glabrous. Fruit up to 8 cm. in diameter, otherwise
similar to *X. granatum*. Seeds 3.5–7.5 cm. long. Fig 21/6.

Kenya.   Kwale District: Diani Forest, Mar. 1973, *Kibuwa* 1203!; Kilifi District: Mnarani, Feb. 1977, *R.B.
& A.J. Faden* 77/553!; Lamu District: Kiunga, Sept. 1956, *Rawlins* 144!
Tanzania.   Tanga District: Sawa, Oct. 1954 (fl.), *Faulkner* 1512! & Feb. 1955 (fr.), *Faulkner* 1565!;
Rufiji District: Mafia I., S. coast, Aug. 1932, *Schlieben* 2671!
Distr.   K 7; T 3, 6; Mozambique, Aldabra, Madagascar, the Mascarenes and throughout most of the
Old World tropics to Australia, Fiji and Tonga, but apparently absent from India except for the
Andamans
Hab.   In coastal scrub just above high-water mark, especially on sandy soil and coral rocks

Syn.   *Carapa moluccensis* Lam., Encycl. Méth. Bot. 1: 621 (1785)

INDEX TO MELIACEAE

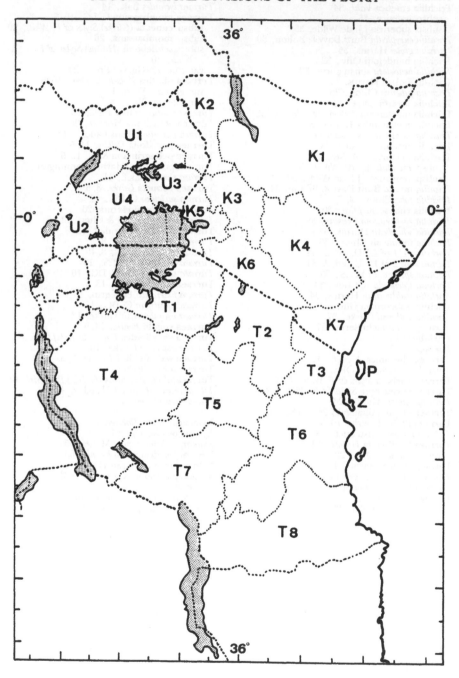

T - #0687 - 101024 - C0 - 244/170/4 - PB - 9789061913566 - Gloss Lamination